Creation, Constellations, and the Cosmos

Norton Simon Museum

Pasadena, California

February 4–June 4, 2001

Contemporary Science and Popular Culture

One Colorado

Pasadena, California

February 4–April 22, 2001

Constructing the Cosmos in the Religious Arts of Asia

Pacific Asia Museum

Pasadena, California

February 4–May 6, 2001

Music of the Spheres

Southwest Chamber Music

Pasadena, California

Concerts at Norton Simon Museum, Armory Center
for the Arts, Colburn School for Performing Arts

September 23, 2000–May 22, 2001

Jets from young stars seen
from the Hubble Space
Telescope, June 6, 1995
NASA/Space Telescope
Science Institute

The Universe

A Convergence *of* Art, Music, *and* Science

Exhibition and Concert Dates

Contemporary Artists and the Cosmos
Armory Center for the Arts
Pasadena, California
February 4–April 22, 2001

Russell Crotty: The Universe from My Backyard
Art Center College of Design
Alyce de Roulet Williamson Gallery, Pasadena, California
February 4–April 22, 2001

The Future of the Universe: Science Fiction Film Festival
California Institute of Technology
Beckman Auditorium and Baxter Lecture Hall, Pasadena, California
January 14–February 27, 2001

Star Struck: One Thousand Years of the Art and Science of Astronomy
The Huntington Library, Art Collections, and Botanical Gardens
San Marino, California
December 1, 2000–May 13, 2001

The Universe

A Convergence *of* Art, Music, *and* Science

Edited by Jay Belloli

With contributions by

Denis Cosgrove

Michelle Deziel and Christine Knoke

Jennifer Gunlock

DeWitt Douglas Kilgore

Dan Lewis

Meher McArthur

Stephen Nowlin

Cornelius Schnauber

Edward Stone

Jeff von der Schmidt

This catalogue is published in conjunction with a series of exhibitions, concerts, and events organized in Pasadena, California. Its publication is made possible by the generous support of Victoria Solaini Baker and the Ralph M. Parsons Foundation.

Published by the Armory Center for the Arts,
145 North Raymond Avenue, Pasadena, CA 91103, U.S.A.

Distributed by:
Reaktion Books Ltd, Publishers
79 Farringdon Road, London ECIM 3JU, United Kingdom
telephone 020 7404 9930 fax 020 7404 9931
e-mail info@reaktionbooks.co.uk
www.reaktionbooks.co.uk

Editors: Karen Jacobson, Sophia Bicos
Designer: Leslie Baker
Production assistant: Jane Teis
Printing: Pace Lithographers, City of Industry, California

ISBN 1-893900-05-3
Library of Congress Card Number: 00-111210

Cover:
Wassily Kandinsky (1866–1944)
Heavy Circles, 1927
Oil on canvas; 22 $\frac{1}{2}$ x 20 $\frac{1}{2}$ in.
Norton Simon Museum, Pasadena, California
The Blue Four Galka Scheyer Collection, 1953

Spine and page 4: Galaxy NGC 7742 from the Hubble Space
Telescope, October 21, 1998, NASA/Space Telescope Science
Institute, Hubble Heritage Team

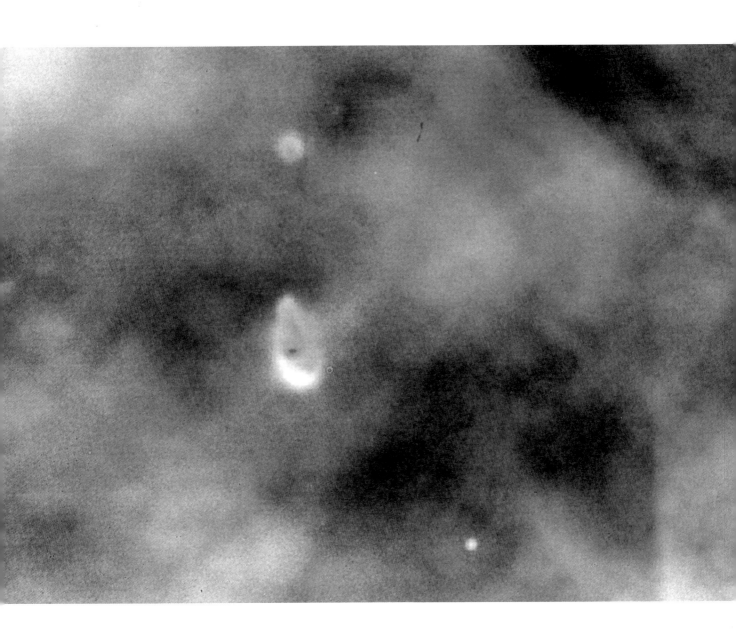

Close-up of
"proplyds" in the
Orion Nebula seen
from the Hubble
Space Telescope,
June 13, 1994
NASA/Space
Telescope Science
Institute

Contents

Giant "twisters" in
the Lagoon Nebula
seen from the Hubble
Space Telescope,
January 22, 1997
NASA/Space
Telescope Science
Institute

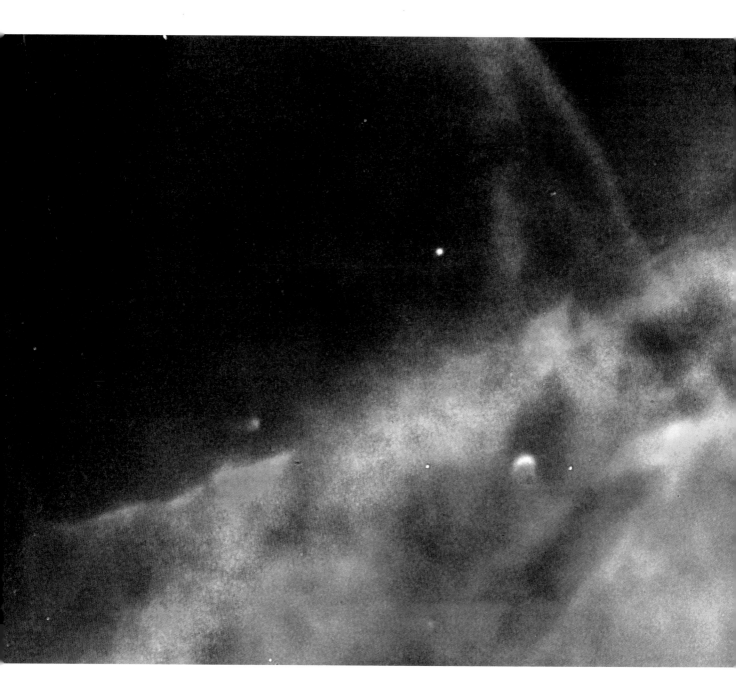

The Orion Nebula seen
from the Hubble Space
Telescope, June 13, 1994
NASA/Space Telescope
Science Institute

Prologue

Pasadena, California, a city of 140,000, is about twelve miles northeast of downtown Los Angeles, at the edge of the San Gabriel Mountains, whose highest peak rises to more than ten thousand feet. Through a combination of geography and history, the Pasadena area has been crucial to the development of the theory of the universe and to its exploration.

The high mountains and sunny, dry climate provide a perfect location for an observatory. At the beginning of the twentieth century the noted astronomer George Ellery Hale moved from the Chicago area (where he founded the respected Yerkes Observatory) to build a new observatory atop Mount Wilson, a six-thousand-foot peak above Pasadena. There he oversaw the building of a one-hundred-inch reflector telescope, then the largest in the world. Hale quickly became involved in the educational and cultural life of the community and was the pivotal figure in making the California Institute of Technology (or Caltech) the world-renowned educational institution it is today. He was also home to one of the key individuals who encouraged the development of the Huntington Library, Art Galleries, and Botanical Gardens, an important scholarly research facility and home to one of the greatest collections of eighteenth-century British art in the United States. Through Hale's efforts (although he did not live to see it completed), an even larger reflector telescope, two hundred inches in diameter, was planned for Mount Palomar, north of San Diego.

The institutions Hale created had an even greater impact than he could have imagined. At the Mount Wilson Observatory in the 1920s, astronomer Edwin Hubble developed the modern conception of the universe and the theory of its origin and expansion. Before Hubble's discovery that a number of objects identified as nebulae (large areas of interstellar gas) were galaxies like our own Milky Way, the universe was believed to be infinitely smaller than we now understand it to be.

The California Institute of Technology became one of the most important institutions in the world for the training of astronomers and a site of significant developments in jet and rocket propulsion. Experiments in propulsion at Caltech led to the creation of the nearby Jet Propulsion Laboratory (JPL), which in the late 1950s led the new era of space exploration with the launch of the Earth satellite *Explorer 1*. JPL became part of the U.S. National Aeronautics and Space Administration (NASA) and the most important center in the world for the exploration of the solar system.

The twentieth century was the most significant period of exploration in history since the late fifteenth and early sixteenth centuries, when Columbus sailed across the Atlantic and other explorers circumnavigated the globe. As we look forward to a new century of discovery, it is important to look back at the past one hundred years, in which the Pasadena area played such a prominent role, in significant part because of George Ellery Hale's vision. Hale was also a bridge between the science and the arts, and *The Universe* project would be inconceivable without the legacy he established.

Jay Belloli
Director of Gallery Programs
Armory Center for the Arts

Preface and Acknowledgments

The Universe is a program of exhibitions and concerts intended to explore humankind's understanding of its place in the universe through an interdisciplinary approach combining the histories of science, art, and music. At the beginning of the new millennium, Pasadena is home to an extraordinary constellation of world-class scientific and space research institutions (California Institute of Technology, Jet Propulsion Laboratory, Mount Wilson Observatory); internationally recognized cultural institutions, museums, and music organizations (Huntington Library, Art Collections, and Botanical Gardens; Norton Simon Museum; Pacific Asia Museum; and Southwest Chamber Music); and noted art education facilities with exhibition programs (Armory Center for the Arts and Art Center College of Design). There is also a retail marketplace in Pasadena, One Colorado, which will make *The Universe* available to a larger public, giving passersby an unexpected introduction to the beauty and challenge of space.

The Universe project itself is a multimedia, multicultural exploration of the cosmos, including exhibitions, performances, art workshops for children and adults, panel discussions, films, seminars, and gallery talks. Chronologically, the project begins at the Huntington, where the exhibition *Star Struck: One Thousand Years of the Art and Science of Astronomy* examines Western civilization's concepts of the cosmos over time through manuscripts, rare books, maps, and photographs. The Pacific Asia Museum explores these ideas in another cultural context in *Constructing the Cosmos in the Religious Arts of Asia*, which reveals the way Asian-born religions have depicted the universe over the centuries. The Norton Simon Museum's exhibition, *Creation, Constellations, and the Cosmos*, explores how artists have defined spiritual connections with the cosmos by highlighting works from its collection of European and Asian art. Southwest Chamber Music also conducts a historical survey of humankind's approaches to the cosmos with *Music of the Spheres*, a concert series with music from the medieval period through the contemporary era.

The relationship between a single human and an entire universe is examined at Art Center's Alyce de Roulet Williamson Gallery in *The Universe from My Backyard*, a showing of two- and three-dimensional drawings by Los Angeles–area artist and amateur astronomer Russell Crotty. At the Armory Center for the Arts, *Contemporary Artists and the Cosmos* comprises works by artists inspired by astronomical photographs, modern scientific theories, and alternate concepts of our place in the universe. One Colorado crosses the boundaries of art and science with an exhibition of photographs from the Hubble Space Telescope and a series of interactive workshops programmed by NASA's Telescopes in Education. At the California Institute of Technology, *The Future of the Universe* is a series of science fiction films, panel discussions, and seminars that explore imaginative and artistic depictions of space and science of the future.

At a time of increasing awareness of the interconnectedness of scientific and artistic disciplines, *The Universe* attempts to give further impetus to this process, exploring some of the ways human beings have attempted to apprehend the vastness and complexity of the cosmos.

*

A series of successful collaborative exhibitions between the Armory's gallery and Art Center's Williamson Gallery in the mid-1990s led in 1999 to the five-institution project *Radical Past: Contemporary Art and Music in Pasadena, 1960–1974*, in which the Armory Center for the Arts, Art Center College of Design, Norton Simon Museum, One Colorado, and Southwest Chamber Music came together to celebrate the "golden age" of the Pasadena Art Museum, its Encounters music series, and artists who lived in Pasadena during the period. This collaboration resulted in unprecedented attendance and critical acclaim for the institutions involved. At the curators' breakfast meeting after *Radical Past's* opening, discussion focused on possible topics for another collaborative exhibition. After a number of ideas generated lukewarm response, Jeff von der Schmidt of Southwest Chamber Music reminded the Armory's Jay Belloli of the exhibition *A Photographic History of the Universe*, which Belloli had organized on a freelance basis several years earlier with encouragement from Stephen Nowlin at Art Center, where it was to have been shown. That interchange was the basis for the extensive collaboration that has developed.

With extraordinary generosity, California Council for the Humanities, the Ralph M. Parsons Foundation, City of Pasadena Cultural Affairs Division and Pasadena Arts Commission (through its Arts and Culture Grants Program), and Victoria Solaini Baker funded the shared program expenses. The exhibitions and programs at the individual institutions have been made possible by the following funders: at the Armory, Victoria Solaini Baker, Los Angeles County Arts Commission, Pasadena Art Alliance, Virginia Steele Scott Foundation, and the Wallace Reader's Digest Funds; at the Huntington, The Boeing Company, Ida Hull Lloyd Crotty Foundation in honor and memory of Edwin and Grace Hubble, Judith and Bryant Danner, GenCorp Foundation-Aerojet, IBM Corporation, the W. M. Keck Foundation, Elise Mudd Marvin, Douglas and Elizabeth Nickerson, Research Corporation, Warren and Katherine Schlinger Foundation, The H. Russel Smith Foundation, the Smithsonian Institution Libraries, and Wells Fargo; at the Pacific Asia Museum, anonymous sponsors and Sam Fogg, London; at Southwest Chamber Music, Los Angeles County Arts Commission, the W. M. Keck Foundation, Ralph M. Parsons Foundation, and the Weingart Foundation.

At the Armory, the staff, board, and Gallery Committee have been extremely supportive of the exhibition, particularly Executive Director Elisa Greben Crystal and Director of Development David Spiro. Linda Centell, gallery program coordinator, and Jennifer Gunlock, gallery assistant, have given essential assistance to the project, as has the installation staff. At Art Center, President Richard Koshalek has encouraged collaborative efforts among Pasadena institutions. The installation has been expertly accomplished by Associate Curator Julian Goldwhite, and gratitude is expressed to Shoshona Wayne Gallery and the collectors who have generously lent works by Russell Crotty to the exhibition. At the California Institute of Technology, President David Baltimore immediately responded to the idea of the exhibition, and Professor Robert A. Rosenstone, chair of the Institute Committee on Art, along with Denise Nelson Nash, director of public events, worked together to develop the interdisciplinary project *The Future of the Universe*, with the assistance of Cara Stemen and Mary Herrera. At the Huntington, W. M. Keck Director of Research Roy Ritchie initiated a separate exhibition and agreed to associate it with this collaborative project. Ronald Brashear, curator of rare books, Dibner Library of History of Science and Technology, Smithsonian Institution Libraries, served as cocurator of the exhibition, and Catherine Babcock, director of communications, and Peggy Spear, major gifts director, have been of great assistance. At the Norton Simon Museum, Director of Art Sara Campbell has been a consistent advocate for the exhibition. Registrar Andrea Clark, Administrative Assistant Kimberly Gilhooly, Assistant Registrar Nicole Hungerford, Curator Gloria Williams, Director of Education Nancy Gubin, Photographer Toni Dolinski, Designer Lilli Colton, Conservator Cara Varnell, and Brian Regan and the installation staff have provided essential help. We are grateful to One Colorado for its magnanimous support of this project and to Gil Clark at NASA's Telescopes in Education for programming a new series of workshops for One Colorado. At Pacific Asia Museum, President of the Board Pat House has provided consistent encouragement of the project, as has Director David Kamansky. Southwest Chamber Music Executive Director Jan Karlin has worked tirelessly on the funding and marketing of the project, and Program Director Inka Bujalski and Development Associate Laura Greene have aided in the production of events. Jonathan Glus, director of the Arts Division for the City of Pasadena, has been a steadfast and enthusiastic advocate for the exhibitions and concerts.

We are grateful to catalogue authors Denis Cosgrove, De Witt Douglas Kilgore, Cornelius Schnauber, and Dr. Edward Stone for their knowledgeable and insightful essays. Karen Jacobson undertook the editing of the catalogue texts, while Armory Marketing Manager Sophia Bicos reviewed early drafts of the texts and coordinated the project. Leslie Baker, with the production assistance of Jane Teis, has designed a beautiful catalogue and invitation. This publication was printed by Pace Lithographers, City of Industry, California, under the care of Carl Bennitt. Robin Ireland, marketing associate for Southwest Chamber Music, provided her expertise in the design of the initial press information and exhibition banners. Many people and institutions have provided photographs for the catalogue, and the Huntington has been particularly generous in this regard. Museums Without Walls assisted in the initial publicity for *The Universe*. We express our sincere appreciation to the museums, collectors, and galleries who lent works of art to the exhibition and provided materials for the catalogue. A number of artists and composers have been extremely generous with their time and assistance, and we are grateful for their contributions.

The Universe attempts to demonstrate what can be accomplished when a number of organizations with special areas of expertise explore a theme that has fascinated humankind since the beginning of history. It is our hope that the project, along with this publication, provokes thought and discussion.

Jay Belloli, Michelle Deziel, Christine Knoke,
Dan Lewis, Meher McArthur,
Denise Nelson Nash, Stephen Nowlin,
Robert Rosenstone, Kate Strauss,
Jeff von der Schmidt
Curatorial Committee

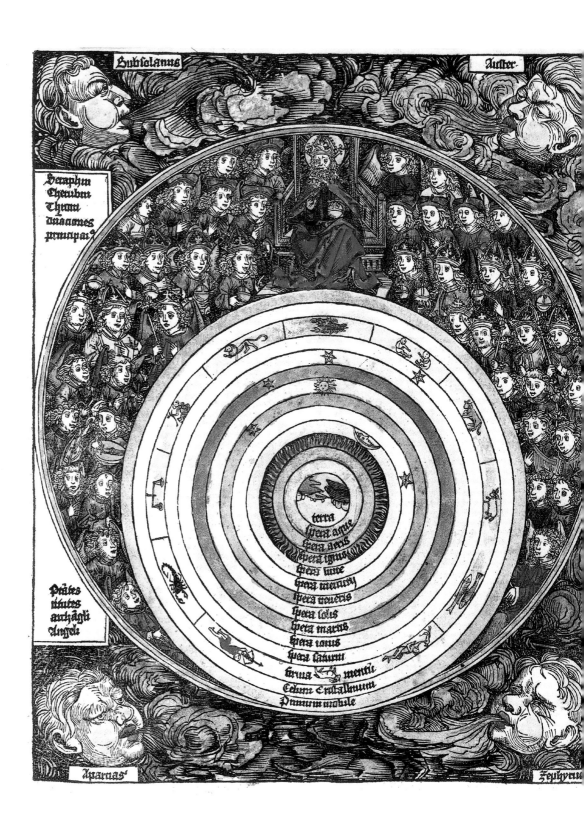

Cosmology *and* Cosmography

1450–1650

The Greek word *kosmos* denotes any harmoniously ordered, fabricated object. When applied to the physical universe, it maintains these associations of conscious fabrication and beauty. Conscious creation brings the idea of cosmos within the scope of religious belief and theological speculation, while the connection with beauty (echoed in today's word *cosmetics*) associates it with ideas of form and order conveyed in mathematics, proportion, and music. The years between 1450 and 1700 in the West are associated with profound changes in religious belief and practice and with dramatic development in the arts. Both were intimately connected with changing Western cosmology.

Every culture has its distinctive cosmology, its explanation of the origins and structure of the universe, and of the place of human life within it. Cosmologies typically incorporate both a temporal account of the origin and evolution of the world, or *cosmogony*, and a graphic description of its spatial structure, form, and content, or *cosmography*. Time and space, always conceptually interdependent, thus find distinct expressions in verbal narrative and graphic description of the universe as an ordered whole. Western culture long took its cosmogony from the Hebrew Bible, whose opening book, Genesis, narrates God's six-day creation of the universe, the earth, and all forms of life. To this, Christianity appended the eschatology of redemption (see fig. 3) and apocalypse to complete the history of creation. Judeo-Christian cosmogony was not consistently challenged until the development of geology's "deep time" in the first decades of the nineteenth century. The West's cosmography, its map of cosmic order, originated in Greek science, above all in Aristotle's theories of the elements, their interrelations and spatial distributions, and in the empirical observations made by classical astronomers and geographers. These were synthesized by Ptolemy in the first

Fig. 1
Hartmann Schedel
Nuremberg Chronicle, 1493
An image of the geocentric Ptolemaic cosmos, showing the four elements, seven planetary spheres, firmament or zone of fixed stars, crystalline sphere, and primum mobile. Beyond are the nine ranks of angels and the creator.
Collection The Henry E. Huntington Library and Art Gallery, San Marino, California

century of the Christian era. It was Aristotelian and Ptolemaic science that was so radically challenged during the early modern period. The challenge was at once philosophical—or, rather, theological—and empirical. As explanations of the place of human life within creation, cosmologies tend always to be teleological and ethnocentric, making the culture in which they originate the purpose of creation, a "chosen people." Similarly, they locate that people at the center of created space. Western cosmology conventionally shared these characteristics, placing Earth at the center of the universe and Jerusalem at the center of Earth. Upheaval of the received world picture, especially after 1500, initiated processes that have simultaneously decentered and universalized Western cosmology. Geocentrism gave way to heliocentrism and ultimately to a universe with no center; imperfection was admitted in the form and motion of physical bodies, both within the heavens and on the Earth's surface; celestial space became unbounded.

Medieval cosmology, derived from Aristotelian physics, pictured a geocentric world machine, filled with matter. The four "corruptible," or changeable, elements—earth, water, air, and fire—composed the sublunar spheres, where linear motion prevailed. Earth and water made up the globe, at whose grossest material depths were the zones of hell, so powerfully described in Dante's *Divine Comedy*. A "meteorological" zone of air and fire extended between the global surface and the lunar sphere, including clouds and all forms of precipitation, winds, and climatic phenomena such as lightning and the aurora. Meteors, comets, and shooting stars, closely observed for prognostication and astronomical science, were also believed to occur within this zone. Beyond were the incorruptible, celestial spheres, within which the seven planets revolved: Moon, Mercury, Venus, Sun, Mars, Jupiter, and Saturn. These spheres were filled with the fifth element, ether, and characterized by uniform circular motion. The material universe was closed by the sphere of the fixed stars, including the constellations of the zodiac, through which the Sun moved over the course of the year. Closely studied, these movements determined the calendar in a predominantly rural society (see fig. 1).

Fig. 2
A British reproduction of an Arabic cosmological astrolabe originally designed for prognostication, c. 1880
Courtesy of the Archives, California Institute of Technology, Pasadena, California

Fig. 3
Book of Hours,
France, early fifteenth
century
Angels of the
heavenly host enter
the celestial sphere to
announce to the
shepherds the birth of
Christ and thus
universal redemption.
Western cosmology
was long based on
the Judeo-Christian
narrative.
Collection The Henry
E. Huntington
Library and Art
Gallery, San Marino,
California

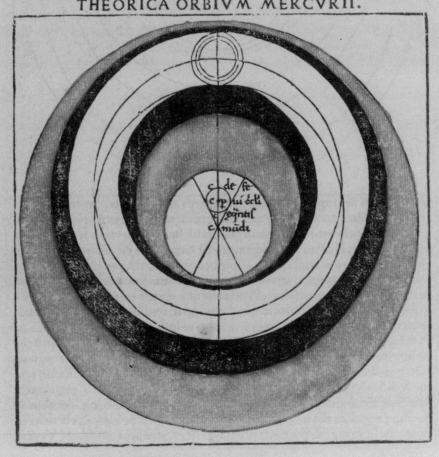

Between the material, visible universe and the supercelestial realm of God, the angelic hosts (themselves sometimes arranged in nine choirs to reflect the material universe), and the elect, two other, invisible zones were sometimes theorized. These reflected varying opinions on the structure and movement of the heavens. These spheres on the outer edge of the fixed stars could help account for observed movement in the firmament, including daily motion from east to west and the precession of the equinoxes. The most widely read medieval works on the spheres, by Sacrobosco and Albert Magnus, suggested a crystalline sphere enclosing the fixed stars, which represented the "waters above the heavens" mentioned in Genesis, and a tenth sphere, or primum mobile, which accounted for the rotation of the whole world machine. The diagram printed in Hartmann Schedel's *Nuremberg Chronicle* of 1493 beautifully illustrates the full extent of this medieval cosmos (see fig. 1).

Medieval scholastics accepted faith and reason as distinct ways of knowing and gave natural philosophy the right to explain physical phenomena according to the laws of nature without recourse to theological argument. Fifteenth-century humanists challenged this principle and blurred the boundaries of reason and faith as distinct epistemologies, appropriate to the material and spiritual worlds, respectively. They demanded that natural philosophy be reconciled with Christian doctrine. Together with new observations and discoveries in terrestrial and celestial space, this gave a new significance to long-recognized inconsistencies and contradictions within medieval cosmology. Theologically, for example, the Aristotelian claims that the cosmos was eternal and that the soul was material and mortal clashed with Christian dogma. Plato, whose work was translated into Latin in the 1460s by the Florentine humanist Marsilio Ficino, offered a response to this problem. His account of cosmic creation in *Timaeus* seemed consonant with Genesis, while the Neoplatonic idea of the soul's ascent through the material spheres to harmony with the divine implied its immortality and chimed with Christian faith. Protestant concentration on salvation by faith, its emphasis on textual exegesis of the newly printed Bible, and belief that God's providential plan was revealed also in nature's book, would all reinforce the metaphysical significance of cosmological questions during the Reformation, in the sixteenth century.

Fig. 5
Nicolaus Copernicus
De revolutionibus
orbium celestium,
1543
Copernicus's
diagrammatic
illustration of
heliocentricity
Collection The Henry
E. Huntington
Library and Art
Gallery, San Marino,
California

Visible irregularities in planetary size and circular motion raised more material questions about the Aristotelian-Ptolemaic cosmos. Observed movements of the planets could be reconciled with the principle of circularity only by means of sophisticated geometrical and mathematical adjustments: the equant and epicycles, which allowed the planets to move eccentrically around the Earth while preserving the perfect circularity of their containing spheres. In southern Germany, the fifteenth-century humanist Georg Peuerbach and his pupil Regiomontanus devised new theories of planetary circulation and crafted new instruments for astronomical observation in response to these problems. The latter completed his tutor's *Epitome* of Ptolemy's great astronomical work *Almagest* and printed *Ephemerides*, or tables of planetary positions, for the years 1475 to 1506. He also printed Peuerbach's *Theoricae novae planetarum*, with its widely reproduced diagrams of planetary movement (see fig. 4). Although he died before the age of thirty, in 1476, Regiomontanus's observational, mathematical, and philological work in relating the elemental spheres and, above all, the circulation of his and other cosmological works in print provided the foundations upon which sixteenth-century astronomy would be built.

In the elemental world, the obvious failure of the water sphere fully to encompass that of earth and the irregular geographical distribution of these two elements offered another challenge to theoretical cosmology. Here too the medieval response leaned toward eccentricity, arguing that the protrusion of the earthly sphere through that of water because their centers were differently located explained the existence and form of a circular world island composed of the three continents and surrounded by ocean. Maritime discovery in the three decades between Vasco da Gama's rounding of the Cape of Good Hope and Magellan's circumnavigation would utterly transform the known distribution of earth and water, revealing a larger, more aqueous, and more geographically diverse globe than Aristotle had theorized or Ptolemy had described. On maps and globes, sixteenth-century cosmographers sought to maintain the balance and symmetry of the Aristotelian elements while mapping an increasingly asymmetrical sphere, for example, retaining a balancing southern continent, Terra Australis Incognita, well into the seventeenth century.

Copernicus's radical reconfiguration of geocentric cosmology grew out of debates over the number, location, and circulation of the planetary spheres necessary to maintain the hypothesis of cosmic order. *De revolutionibus orbium coelestium* (1543) was illustrated by a simple, powerful redrawing of the conventional image of the cosmos which centered it upon the Sun, with the Earth and its lunar satellite placed in a third sphere (see fig. 5). Copernican heliocentrism was the outcome of traditional forms of astronomical reasoning rather than new astronomical observation, and its computational complexities seemed to most readers to offer little improvement over Ptolemy's system, so that arguments for heliocentricity convinced no more than a handful of sixteenth-century thinkers. Interest increased following the supernova of

Fig. 6
Tycho Brahe's
observatory on the
Danish island of
Hven (1580), the
most sophisticated
European
astronomical center
before the advent of
the telescope.
Collection The Henry
E. Huntington
Library and Art
Gallery, San Marino,
California

ORTHOGRAPHIA PRÆCIPVÆ DOMVS ARCIS VRANIBVRGI
in Infula Porthmi Danici Venufia, *vulgo* Huenna,   Aftronomiæ inftaurandæ gratia, circa annum MDLXXX,
à TYCHONE BRAHE   exædificatæ.

1572 and Pope Gregory XIII's calendar reform of 1584, which brought renewed
attention to astronomical observation. But the observational advances of Tycho Brahe
and Johannes Kepler remained constrained by their instruments, although Tycho's
observatory at Hven, Denmark (see fig. 6), was the most sophisticated in Europe and
led to his hybrid geocentric cosmology, with the two inner planets circling the Sun.
By 1600 Tychonian, Ptolemaic, and Copernican models were becoming familiar
alternative images of the cosmos, widely reproduced and promoting intense interest in
the world machine among educated Europeans. It was Galileo's use of the telescope,
however, invented about 1608, that empirically undermined the perfection of
Aristotelian celestial cosmography, revealing corrugations on the lunar sphere, moons
revolving around Jupiter, and imperfections on the surface of the Sun. And it was
Newton's demonstration of universal gravity that finally demolished Aristotle's
principles of cosmic motion and introduced a new cosmology.

A perfit defcription of the Cæleftiall Orbes,

according to the moſt auncient doctrine of the
Pythagoreans. &c.

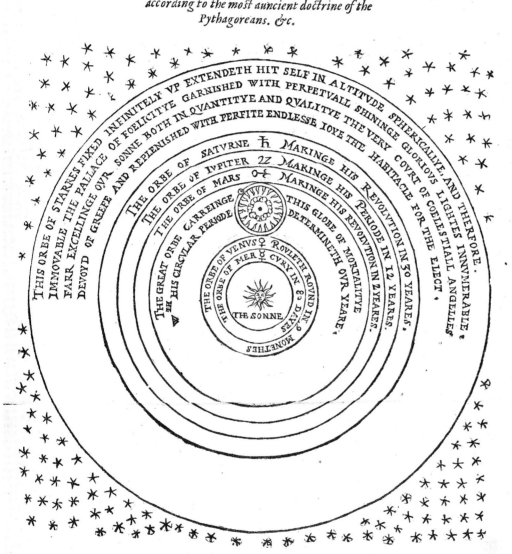

M. .i.

A PER-

Fig. 7
Thomas Digges
A Perfit Description
of the Caelestiall
Orbes, 1576
Thomas Digges's
book, published and
illustrated as an
appendix to his
father's perpetual
almanac
(Leonard Digges,
A prognostication
everlastinge . . .
[London, 1576]),
paraphrased
Copernicus's first
book and illustrated
its implications of an
infinite cosmos in a
heliocentric map that
breaks the sphere of
the fixed stars.
Collection The Henry
E. Huntington
Library and Art
Gallery, San Marino,
California

Fig. 8
Johannes Kepler
*Mysterium
cosmographicum*,
1597, ed. 1621
Kepler's illustration
of the five platonic
solids in his
*Mysterium
cosmographicum*
related the
geometrical forms to
the measured
planetary orbits of
Mercury, Venus,
Earth, Mars, and
Jupiter in a
heliocentric cosmos.
The outermost sphere
is Saturn's.
Courtesy of the
Archives, California
Institute of
Technology,
Pasadena, California

Sebastian Münster's hugely popular *Cosmographia* of 1544, structured, like Schedel's *Chronicle*, as an encyclopedia of cosmography following the narrative of Christian history, appeared the year following Copernicus's *De revolutionibus*, Vesalius's *De humani corporis fabrica libri septem*, and Niccolo Tartaglia's first vernacular translation of Euclid's *Elements*. Münster's cosmography was a humanist encyclopedia, illustrating in words and pictures the universal majesty of God's creation, as its frontispiece and illustrations of *mirabilia*, or fantastic marvels from newly discovered places, make clear. By the mid-sixteenth century *cosmos* had become a familiar trope for rational observation and description of natural phenomena, from the scale of the world machine to that of the human microcosm, and metaphysical connections were frequently drawn between these different scales. Catholics and Protestants alike had long since abandoned the scholastic separation of faith and reason, and each required natural philosophy to submit to religious doctrine. The cosmological concept of a providentially ordered and harmonious "world machine," especially when connected to Neoplatonic ideas of the soul's ascent toward divine love, offered for many a retreat from religious strife and a possible point of doctrinal resolution. But if pietism sustained cosmology's harmony, the unrelenting flow of observational data from navigation and systematic celestial observation undermined its synoptic intent of unifying earth and heavens.

By the end of the sixteenth century, terrestrial cosmography, which since the encyclopedias and universal histories of the Middle Ages had sought to give unity and structure to the geographical patterns of the globe's surface, had been radically disrupted. The Americas represented a fourth part of the world, with a population quite separate from Noah's progeny; the Torrid Zone had been shown to be habitable; the antipodes were populated; Jerusalem was not the center of a single land hemisphere. A full explanation of terrestrial magnetism remained elusive, however, and accurate determination of longitude at sea would not become possible until the late eighteenth century. Thus the true geography of the vast and varied phenomena making up the terraqueous globe remained vague, undermining the value of the cartographic grid of latitude and longitude as an instrument of cosmographic order.

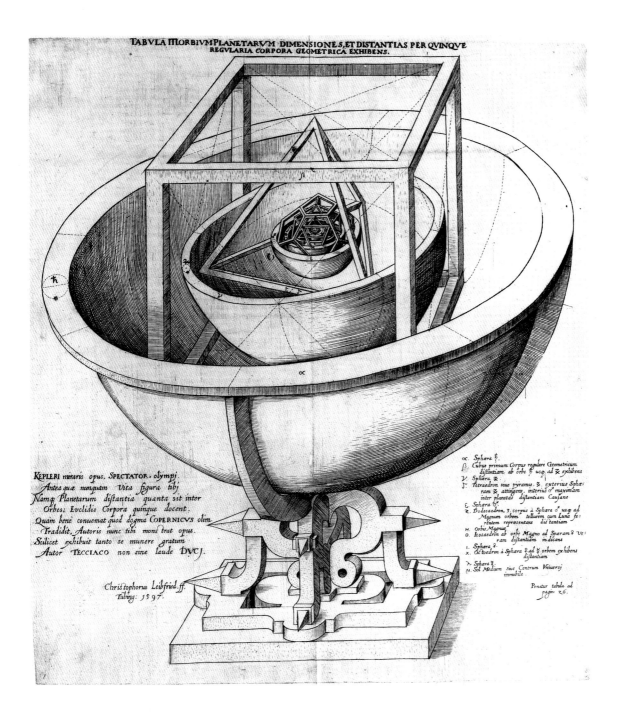

TABVLA MORBIVM PLANETARVM DIMENSIONES, ET DISTANTIAS PER QVINQVE
REGVLARIA CORPORA GEOMETRICA EXHIBENS.

KEPLERI miraris opus, SPECTATOR. olympi,

Antea quæ numquam Vita figura tibi,
Namq; Planetarum distantia quanta sit inter
Orbes: Euclidis Corpora quinque docent.
Quam bene conuemat quod dogma COPERNICVS olim
Tradidit, Autoris nunc tibi monstrat opus.
Scilicet exhibuit tanto se munere gratum
Autor TECCIACO non sine laude DVCI.

Christophorus Leibfried. ff.
Tubing: 1597.

α. Sphæra ♄.
β. Cubus primum Corpus regulare Geometricum
     distantiam ab orbe ♄ usq; ad ♃ exhibens
ϒ. Sphæra ♃.
δ. Tetraedron siue pyramis ♃. exterius Sphæ:
     ram ♃ attingens, interius ♂ maximam
     inter planetas distantiam Causans
ε. Sphæra ♂.
ζ. Dodecaedron, ♂ corpus à Sphæra ♂ usq; ad
     Magnum orbem tellurem cum Luna fe:
     rentem repræsentans distantiam
H. Orbis Magnus
θ. Icosaedron ab orbe Magno ad Spæram ♀ Ve:
     ram distantiam indicans
I. Sphæra ♀.
χ. Octoedron à Sphæra ♀ ad ☿ orbem exhibens
     distantiam
λ. Sphæra ☿.
μ. Sol Medium siue Centrum Vniuersi
     immobile

Ponatur tabula ad
pagin: 26.

The late sixteenth century also witnessed an emerging dispute over the universal extension of the world machine itself. In his 1576 world system, one of the earliest to illustrate Copernicus's argument, Thomas Digges's notation, if not his drawing, embraced the full implication of heliocentrism by combining fixed stars and the empyrean (see fig. 7). To account for the distances and speeds of planetary revolution in a heliocentric cosmos, its dimensions must be stretched almost to infinity and a huge void must exist between planets and stars, violating Aristotle's principle of plenitude. Heliocentrism also breaks the contiguity and ultimately the very existence of the Ptolemaic spheres, as Kepler's *Mysterium cosmographicum* (1596) made absolutely apparent. Kepler's work signaled a shift in cosmological reasoning, away from the Euclidian geometry appropriate to a fixed and finite cosmos, toward the Archimedian mathematics appropriate to studying motion and attraction between bodies in infinite space. But Kepler was no materialist; his metaphysical speculations are well known, and his Platonism was reinforced by his discovery that the relative dimensions of the planetary orbits could be related to the forms of the five Platonic solids (see fig. 8).

By 1620 the telescope and microscope were generating volumes of new astronomical observation, revealing previously invisible structures within elemental matter. The claims of autopsy (seeing for oneself) over other forms of authority, and of experience/experiment over rhetoric in scientific discourse, were increasingly asserted. These advances intensified questions of verifying cosmological observation. The longitude problem remained unsolved, while the phenomena revealed by optical instruments could be made public only by means of graphic images. Questions of vision and the veracity of images underlay metaphysical disputes between Kepler and Robert Fludd over the structure of the cosmos. Celestial phenomena, such as the lunar craters mapped by Kepler and Jupiter's moons, observed by Galileo, might challenge faith in the perfection and harmony of an Aristotelian-Ptolemaic cosmos, but they by no means swept it away. Jesuit astronomers such as Christopher Scheiner, Clavius Fabius, and Giambattista Riccioli used the new instruments to map celestial phenomena within the conventional Ptolemaic frame. Galileo's famous sunspot images of 1613, traced by the lens directly onto paper, lent support to the idea that mechanization of the image might guarantee its truth (see fig. 9). In all ways, therefore, experiment and image making, although socially disparaged as "mechanical" rather than "liberal" arts, assumed an increasingly significant role within natural philosophy.

Sir Isaac Newton's *Principia* of 1687 established a law of attraction between bodies in free space—universal gravitation— whereby every particle of matter in the universe attracts every other particle with a force that varies directly as the product of their masses, and inversely as the square of the distance between them. This universal law applies to spherical shells of constant density, indicating that no external force is required to maintain the patterns and motions of the planets. The Aristotelian-Ptolemaic cosmology was effectively laid to rest, and the foundations of a wholly secular cosmology

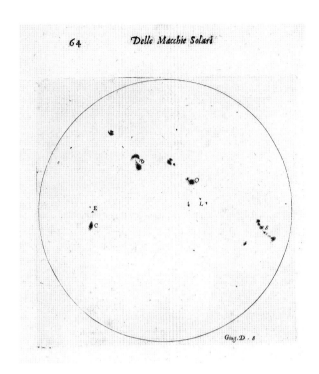

Fig. 9
Galileo's image of sunspots burned directly onto paper through his telescope, 1510
Collection The Henry E. Huntington Library and Art Gallery, San Marino, California

put in place. But we should recall that the same decade saw the zenith of absolute sovereignty in the rule of Louis XIV of France, a monarch whose life and reign were scripted through the discourse of cosmology, realized in the design of his palace at Versailles. The Sun King placed himself at the territorial and symbolic center of a political universe, his image endlessly reflected in the Hall of Mirrors and carried to the limits of space along the axes that radiated from his court. At either side Louis had placed the largest globes then manufactured: one mapped his earthly empire and the geographical discoveries made by France; the other showed the pattern of planets and stars in the heavens at the moment of the king's nativity. Then, as now, cosmology was as much a social discourse as a scientific one.

Denis Cosgrove is Von Humboldt Professor of Geography at the University of California, Los Angeles.

The Melody
of Numbers

If I were not a physicist, I would be a musician.
Albert Einstein

The Harmony of the Universe has been a marriage of music and mathematics since Greek antiquity. It was Pythagoras who began the study of abstract numbers in sixth century Greece and forged the genesis of musical theory with his anvil yard observation of the hammering vibrations of fourths, fifths, and octaves. Today, however, it has become as difficult to imagine the interpenetrating of music and mathematics as it is to hear the Music of the Spheres.

My purpose, from the vantage point of the beginning of the Third Millennium and *The Universe* festival here in Pasadena, is to begin to refresh the tradition of music and mathematics. Through a cultural self-portrait of ideas that blends Greek Antiquity, Medieval, and Renaissance thought with our own era, I hope to encourage a new sensibility about both subjects. Because the ultimate question could well be—were past explanations of the cosmos as satisfying to those eras as ours are to us today?

Is science only meant to subjugate nature, music only an idle distraction? Was Einstein articulating a profound connection between music and science that should be taken as far more than an amusing anecdotal wandering? Is it time to exonerate Schoenberg from the scorn of having developed a "mathematical" dodecaphonic music?

Fig. 10 (opposite)
Urania, Astronomy, and Ptolemy, from the *Spheara mundi* of Johannes de Sacrobosco (1510)
The Huntington Library Rare Book Collection

Fig. 11 (left)
The Seven Liberal Arts, from Martianus Capella's *Marriage of Mercury and Philology*
The Huntington Library Rare Book Collection

Opus

Martiani Capelle de Nuptijs
Philologie z Mercurij libri duo
De grammatica.
De dialectica.
De rhetorica.
De geometri .
De arithmetica.
De astronomia.
De musica libri septem.

A curious thing happened in 1653. While trying to reach an "ingenious reader" of 17th century London with a catchy preface aimed at boosting sales, an uncredited English translator of Descartes' *Excellent Compendium of Musick: with Necessary and Judicious Animaadversions Thereupon* summed up centuries of theory and practice about music, shedding light on the entire spectrum of education.[1]

And, as for the SUBJECT likewise, wherewith the Rationall Soule of Man is so Pathetically, and by a kinde of occult Magnetisme, Affected, that even the most Rigid and Barbarous have ever Confest it to be the most potent Charme either to Excite, or Compose, the most vehement Passions thereof. That the Sage and Upright Ancients had Musick in so high Estimation, as that, when they would fully Characterise a Learned and Sapient Person, they called him only a Musician: as if to be well skilled in the Concordant and Discordant Proportions of Numbers, were the most perfect Diapason of Virtue and Knowledge. Thus even the best of our Moderne Grammarians, and Philologers, derive the word Musick, as also the Muses, from the Greeke verb that signifies to Explore with Desire: and this, upon no slender Reason; insomuch as the Key that opens the difficult Locke of all Arts and Sciences, must be easily Collected from this Consideration; that to be a Complete Musician (please you, to understand Him to be such, as hath not only Nibbled at, but swallowed the whole Theory of Musick) is required a more than superficial insight into all kinds of Humane Learning. For, He must be a Physiologist; that he may demonstrate the Creation, Nature, Proprieties, and Effects of a Natural Sound. A Philologer, to inquire

1. Rene Descartes, *Excellent Compendium of Musick: with Necessary and Judicious Animaadeversion thereupon* (London: Humphrey Moseley, 1653).

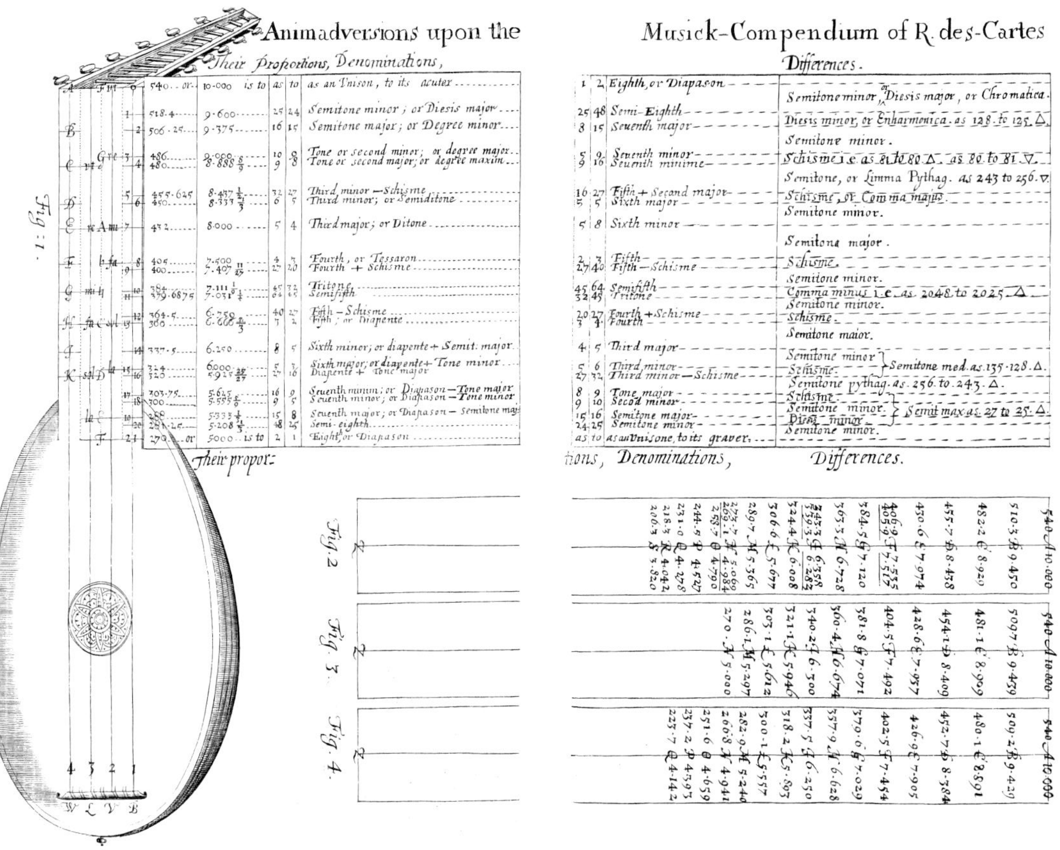

Fig. 12
René Descartes' computations based on the vibrations of the strings of the lute (English edition, 1653)
The Huntington Library Rare Book Collection

into the first Invention, Institution, and succeeding Propagation of an Artificial Sound or Musick. An Arithmetician, to be able to explain the Causes of Motions Harmonical, by Numbers, and declare the Mysteries of the new Algebraical Musick. A Geometrician; to evince, in great variety, the Original of Intervalls Consono-dissonant, by the Geometrical, Algebraical, Mechanical Division of a Monochord. A Poet; to conform his Thoughts, and Words, to the precise Numbers, and distinguish the Euphonie of Vowells and Syllables. A Mechanique; to know the exquisite Structure or Fabrick of all Musical Instruments, Winde, Stringed, or Tympanos alias Pulsatile. A Metallist; to explore the different Contemperations of Grave and Acute toned Metals, in order to the Casting of tuneable Bells, for Chimes, &c. An Anatomist: to satisfie concerning the Manner, and Organs of the Sense of Hearing. A Melothetick; to lay down a demonstrative method for the Composing, or Setting of all tunes, and Ayres. And lastly, He must be so far a Magician, as to excite Wonder, with reducing into Practice the admirable Secrets of Musick: I meane, the Sympathies and Antipathies betwixt Consounds and Dissounds; the Medico-magical Virtues of Harmonious Notes (instanced in the cure of Sauls Melancholy fits, and of the prodigious Venom of the Tarantula, &c.); the Creation of Echoes, whether Monophone, or Polyphonie, i.e. single or Multiplied, requisite to the multiplied Reverberations of Sounds; the Artifice of Otocoustick Tubes for the strengthening, continuation, and remote transvection of weak sounds, and the mitigation of strong; and finally, the Cryptological Musick, whereby the secret Conceptions of the mind may be, by the Language of inarticulate Sounds, communicated to a Friend, at good distance.

This entertaining litany goes a long way in explaining the universal language and attraction of music. Originating in Classical Greek and Roman educational traditions, the unity of the arts and sciences was demonstrated throughout the Medieval and Renaissance eras by the Latin term quadrivium. When combined with the trivium—logic, grammar, and rhetoric—the two created the seven liberal arts. The quadrivium established four essential curricula that embraced a virtual Copernican center of Pythagorean thought: the study of number.

QUADRIVIUM

Arithmetic
Number ITSELF

Music
Number in PROPORTION

Geometry
Number at REST

Astronomy
Number in MOTION

Music provided the justification and proof of scientific theory in the Renaissance. The societal importance and prominence of music originated with Pythagoras, Aristotle and Plato as explained by the Neo-Platonist philosopher Boethius. Boethius, one of the great Latin encyclopedists who revived the intellectual traditions of antiquity in the Middle Ages, articulated three types of music.

Musica Mundana, or Musical Science, was concerned with the proportions found in the distances of the celestial world. Harmonic proportions measured these distances through the vibrations of a monochord. It was believed that each planet emitted a specific tone, thus giving credence to the inaudible "Music of the Spheres."

Musica Humana, or Musical Philosophy, represents music's role as an instrument of Neo-Platonist philosophy. The mathematical proportions of human form weave a continuous web, unifying the physical nature of man with the relationship to his internal chemistry. The amalgamation of these physical and chemical proportions became the pathway to the soul.

Musica Instrumentalis, or Musical Performance, is the only form of the Medieval and Renaissance perception of music that exists today. The resultant proportions of the combination of the audible and inaudible yields the ordered sounds produced by the human voice or musical instruments.

But it was number that exerted the ultimate influence on all the disciplines of the quadrivium. As Heinrich Cornelius Agrippa, a court counselor to Charles the Fifth, wrote in his *On Occult Philosophy* of 1533: "*Severinus Boethius* saith, that all things which were first made by the nature of things in its first Age, seem to be formed by the proportion of numbers, for this was the principal pattern in the mind of the Creator. Hence is borrowed the number of the Elements, hence the courses of times, hence the motion of the Stars, and the revolution of the heavens, and the state of all things subsist by the uniting together of numbers. Numbers therefore are endowed with great and sublime virtues."[2]

The imaginative necessity of number in measuring the vast concepts of nature and the cosmos has been a constant in the equation of humankind. In one of the most apt artistic descriptions of the 20th century world of Einstein and Hubble,

2. Heinrich Cornelius Agrippa, *On Occult Philosophy* (London, 1651), Huntington Library Rare Book Collection.

Ernst Krenek, whose opera *Karl V* deals with the life of Agrippa's king and was the first opera to use Schoenberg's twelve-tone technique, brought the fascination of number full circle with his *Sestina*.[3]

3. Ernst Krenek, *Sestina* (1957), Bärenreiter, Kassel, Germany.

> *As I with measure master sound and time*
> *Shape recedes in unmeasured chance.*
> *The crystal of number releases life's stream*

Number Itself

The great German mathematician Carl Friedrich Gauss had the type of success that school children dream of but rarely experience. Asked to produce the sum of all the numbers from one to a hundred, Gauss answered the question quickly and easily, and without addition: 5050.

What he had observed was that the sum of the first and last numbers, one and a hundred, was a hundred and one. That led him to the next pair, two and ninety-nine, which also yielded the sum of a hundred and one. He quickly reached the conclusion that there had to be fifty pairs of such numbers, all yielding the sum of a hundred and one. The answer had to be 5050.

$$1 + 2 + 3 \ldots 49 + 50 + 51 + 52 \ldots + 98 + 99 + 100$$
$$101$$

$$50 \times 101 = 5050$$

Gauss, the father of modern number theory, has a musical counterpart in Beethoven. In a moment we will look at Beethoven in Gaussian terms. But before that we need to build upon the arithmetic of the quadrivium, represented by Gauss' fascination with number itself, and look at number in proportion, rest and motion.

The Great Monochord:
Musica Mundana

A Rosicrucian eccentric in the eye of history, Robert Fludd was born in 1574. He was recognized as one of the finest doctors of the Renaissance. But through his entire soul and being Fludd was a cosmologist. Though he provoked the ire of Kepler and others, there is no denying the intensity and vision of his pantheistic achievements. His various cosmological and medical books include perhaps the most captivating engravings and imaginative texts of an entire era.

Fig. 13
The Divine Monochord of Robert Fludd
The Huntington Library Rare Book Collection

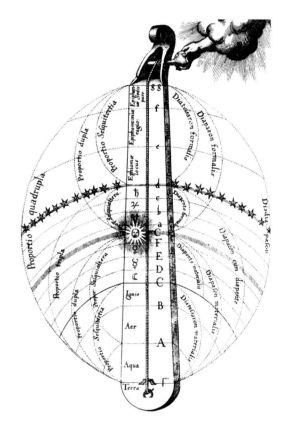

In figure 13, we have *The Divine Monochord* from Fludd's *Utriusque Cosmi Historiæ* of 1617. In the upper right hand corner, the hand of God is tuning the string of the monochord, which will stretch from the double gg down to the final Greek gamma. To the sides of the notes, the smallest distance of the intervals can be found in the shaded area. Proceeding up from the Greek gamma, we travel through the four elements of earth, water, air, and fire. After the elements we reach the seven planets of the Moon, Mercury, Venus, the Sun, Mars, Jupiter, and Saturn. We then proceed to the three reaches of the empyrean realm, the galaxies and nebula of Hubble today, the *cœllum stellatum* of Dante. To the right side of the monochord we have spherical diagrams of the Greek names of each intervallic proportion (*Disdiapason* = double octave; *Diapason* = octave; *Diapente* = fifth; *Diatessaron* = fourth). To the left side we have the numerical proportions of the Greek intervals that appear on the right. String lengths of 1 : 4 result in the sounding of a double octave from the lowest *gamma* to the highest double g, or a *Proportio*

quadrupla. The other relationships arc *Proportio dupla* 1 : 2 and 2 : 4, the octave; *Proportio sesquialtera* 2 : 3, the fifth; and *Proportio sesquitertia* 3 : 4, the fourth. These proportions are the manifestations of the wandering notes of the planets, the Music of the Spheres.

From Antiquity through the Renaissance music provided the proof of numerical and geometric proportion. As we look at Fludd's monochord, we can also understand the conjoining aspects of number at rest, their geometry. The right and left sides of the monochord represent a spherical universe.

Contrary to popular belief, past eras did not firmly believe the earth was flat. Direct observation of the Sun and Moon mitigate against such ideas. The many representations of cosmological charts confirm the cultural acceptance, long before Columbus, that the shape of the earth, the heavens, and the empyrean realm of the celestial stars, was spherical.

In 1854, Bernhard Riemann, the Mozart and Brahms of mathematics rolled into one, delivered one of the most critical lectures in the advancement of non-Euclidean geometry, wandering close to the world of the unified field theory. Straight lines, though boundless, were finite in length. Einstein summed up Riemann's ideas in a single brilliant statement: "Riemann's geometry on an n-dimensional space bears the same relation to Euclidean geometry of an n-dimensional space as the general geometry of curved surfaces bears to the geometry of the plane."

In musical terms, Brahms became the progressive for Arnold Schoenberg. The advances in the 19th century, either in the discovery of non-Euclidean geometry in mathematics or the expansion of harmonic materials in music, would prepare the way for the discoveries of the 20th century.

The spherical representation of the universe remained a constant. The mathematics became the variable. The images in Figures. 1, 4 and 5 can be interpreted,

Fig. 14
The Ptolemaic Universe II of Robert Fludd The Huntington Library Rare Book Collection

and many other examples could be chosen, as archetypal ancestors to the non-Euclidean theories of Riemann, whose conception of space was spherical. That the shape of the sphere may hold the key to the shape of the universe would be the starting point of Einstein's theory of relativity.

In the Middle Ages and Renaissance, the vibrations of the monochord string, which still correspond to the open strings of any violin in use today, confirmed the measurements of mathematics, geometry, and astronomy.

In the 20th century, Einstein would remark, "If I were not a physicist, I would be a musician."

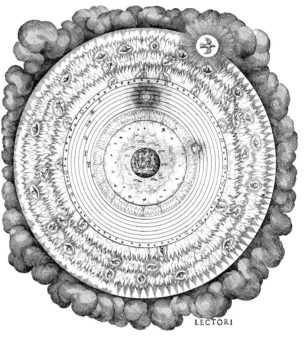

Micro and Macrocosm: Musica Humana

As we have said earlier, Musica Humana represented musical philosophy. No one represented the totality of the human placement in divine order as well as Robert Fludd. On the frontispiece of the *Utriusque Cosmi Historiæ* (fig. 15) Fludd relates the human figure to both micro and macrocosm. In the microcosm, the male figure is surrounded by the twelve signs of the zodiac. The four elements relate to the four humors of Galenic medicine. The macrocosm revolves around the inner sphere of man, and is pulled by Saturn. With reflection and familiarity, one can experience the perpetual revolution of this image: it is not meant to represent a static state.

"The working-out in breadth, length, height, and depth begins in my head...and I hear and see the picture as a whole take shape and stand forth before me

4. Alexander Wheelock Thayer, *The Life of Beethoven* vol. 2 (Princeton: Princeton University Press, 1967), 851.

as though cast in a single piece, so that all that is left is the work of writing it down."[4] Beethoven is reported to have had this conversation with a student while he was at work on the *Ninth Symphony*. Like Gauss, Beethoven was able to intuit solutions that would include both the large and the small at once. With Beethoven the micro and macrocosm of the tonal musical universe exists simultaneously.

Though a unique work in every aspect, Beethoven's *Hammerklavier* sonata demonstrates his ability to build large structures from the same material as the smallest detail. Beethoven's power resides in his exhaustive creativity with the building blocks of music. Harmony and its smallest ingredient, the proportion of the interval, become the basis of the composition.

In the case of the *Hammerklavier*, the harmony is B flat major and the interval is the third. Classical music had perfected the relation of tonic to dominant (*Diapason to Diapente*). But to express a vastly expanded musical universe, Beethoven focuses this grand work on smaller relationships, the interval of the third and the resulting clash of major and minor seconds resulting from this hierarchy.

The form of the first movement is a triple descent through keys related by the third, which results in placing the greatest harmonic clash, between B and B flat, at the heart of movement.[5]

5. Charles Rosen
The Classical Style
(New York:
W.W. Norton, 1971).

B flat major	Primary theme group} Exposition
G major	Secondary groups} Exposition
E flat major	Opening of Development
B major	End of Development
B flat major	Recapitulation
G flat major	Elaboration (Diapente of B major)
B major	Diapason of G flat major
B flat major	Confirmation of Tonic

One can clearly see the large scale scope of the first movement of the *Hammerklavier*. But as one looks at the triple descent of thirds and the resolution of those descents, one begins to understand that, like a static appreciation of the spherical image of Fludd's micro and macrocosm, one does not fully appreciate Beethoven in a linear state.

Music can help understand the shape of space if a linear perception of time is abandoned. Tonality functions around the circle of fifths. As can be seen from the shape of the modulations of the first movement of the *Hammerklavier*, B flat major returns in a rotational way, much like the joining of the sides of a cylinder. This accounts for the recognition one feels when the governing tonic of B flat major comes into the ear. The music *returns* to B flat major; it does not progress to it in a straight line. When this is understood, the universe of music begins to reflect the curvature of space.

It was not until the 20th century that music would realize the continued rotational expansion of its materials, in much the same way that science would come to greater terms with the shape of the natural universe. As Hubble and others observed the retroverse of the universe from our perspective here on earth, Arnold Schoenberg would design the most beautiful musical metaphor for the great cosmological fervor of the 20th century. The Melody of Number was at hand.

Ars Combinatoria

In 1669 the Jesuit priest Athanasius Kircher postulated integers into a hypothetical infinity in his *Ars Magna Scienda* (fig. 16). In *Ars combinatoria*, the fourth chapter of his magnificent tome, Kircher examines in exhaustive detail the meaning of complete potential of material. The chapter ends with a striking design that would highlight the paradox of straight lines creating curved surfaces, centuries in advance of M. C. Escher (fig. 17).

"All phenomena are similar, and none are identical" writes Goethe in his *The Metamorphosis of the Plants*. The belief in the unity of all things—a basic tenet of cosmology—has led humankind to follow its intuition to logical conclusions. But as Einstein pointed out "the incomprehensible thing about the universe is that it is comprehensible". In searching for a unified field theory, Einstein would meet with the same skepticism that greeted the elegant logic of Schoenberg's dodecaphonic universe. The basic idea behind Schoenberg's harmonic innovation is a compendium of the contrapuntal techniques of inversion, retrograde, and retrograde inversion. This Bach-like world would prove to be quite flexible.

ſive COMBINATORIÆ, LIB. IV.

TABULA GENERALIS.

*Ex quâ omnes rerum ſimpliciter commutandarum conjugationes poſsibiles
eruuntur.*

```
 1 | 1. A.
 2 | 2. B.
 3 | 6. C.
 4 | 24. D.
 5 | 120. E.
 6 | 720. F.
 7 | 5040. G.
 8 | 40320. H.
 9 | 362880. I.
10 | 3628800. K.
11 | 39916800. L.
12 | 479001600. M.
13 | 6227020800. N.
14 | 87782912000. O.
15 | 1307674368000. P.
16 | 20922789888000. Q.
17 | 355687428096000. R.
18 | 6402373705728000. S.
19 | 121645100408832000. T.
20 | 2432902008176640000. V.
21 | 51090942171709440000. X.
22 | 1124000727777607680000. Y.
23 | 25852016738884976640000. Z.
24 | 620448401733239439360000.
25 | 15511210043330985984000000.
26 | 403291461126605635584000000.
27 | 10888869450418352160768000000.
28 | 304888344611713860501504000000.
29 | 8841761993739701954543616000000.
30 | 265252859812191058636308480000000.
31 | 8222838654177922817725562880000000.
32 | 263130836933693530167218012160000000.
33 | 8683317618811886495518194401280000000.
34 | 295232799039604140847618609643520000000.
35 | 10333147966386144929666651337523200000000.
36 | 371993326789901217467999448150835200000000.
37 | 13763753091226345046315979581580902400000000.
38 | 523022617466601111760007224100074291200000000.
39 | 20397882081197443358740281739902897346800000000.
40 | 815915283247897734345611269596115893872000000000.
41 | 33452526613163807108334020534407516473520000000000.
42 | 1405006117752879898543002892624454511569188784000000000.
43 | 60415263063373835637051243828513997475117712000000000000.
44 | 265827157478848876808664752845461588905917932800000000000.
45 | 10765999877893217410629451651241194350065976278400000000000.
46 | 49523599438303880008889547759570949401030349088064000000000000.
47 | 2327609173600513604178087446998346218484264071390080000000000000.
48 | 111275240332824653000548197455920618487244675426723840000000000000.
49 | 5474536776308407997026861675340110305874989095909468160000000000000.
50 | 12737268388154203998516343083767005515293749454795473408000000000000000.
```

Artis Magnæ Sciendi,

EPILOGISMUS
Combinationis Linearis.

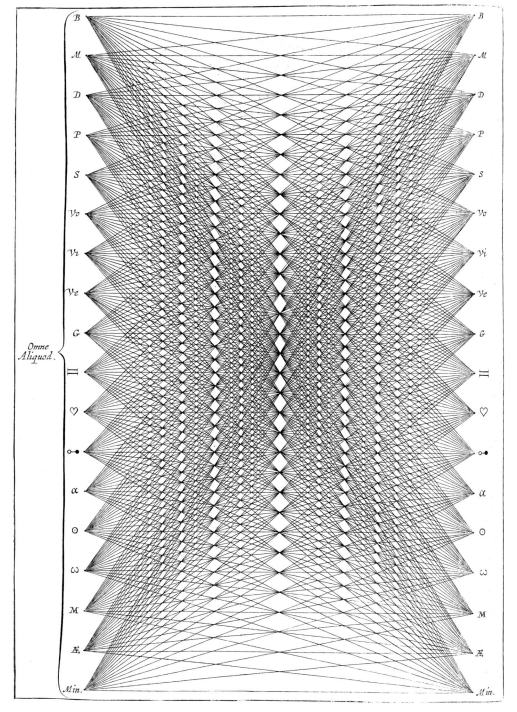

Fig. 17
From the "Ars combinatoria" of Athanasius Kircher's *Ars magna scienda*, 1669
The Huntington Library Rare Book Collection

```
S   A   T   O   R
A   R   E   P   O
T   E   N   E   T
O   P   E   R   A
R   O   T   A   S
```

"Sator the Reaper Keeps the Wheel Turning". This is a palindrome beloved of Anton Webern, one of Schoenberg's most important students. A magic formula from Pompeii, it explains both the philosophy and proportion of the original, inversion, retrograde, and retrograde inversion of the dodecaphonic world. It reads the same forwards, backwards, upwards, and downwards.

As Robert Osserman has pointed out "Looking out in space and looking back in time are one and the same...every galaxy we can observe— our entire retroverse—must have started in the same place as our own galaxy some twenty billion years ago. Said differently, running our universe backward from the current positions and velocities—as close as we can determine them—we find the whole thing collapses together some twenty billion years in the past."[6]

6. Robert Osserman
Poetry of the Universe: A Mathematical Exploration of the Cosmos (New York: Anchor Books, 1995).

Fig. 18
The 100-inch telescope on Mount Wilson, Carnegie Institution of Washington

Edwin Hubble clarified the main point of the Riemannian-Einstein view of the shape of the universe even further. Hubble, working on the top of Mount Wilson with the first 100" telescope (fig. 18), realized that other galaxies are in a recession from our own. The rate of their recession is dependent on their distance. The ratio between their velocity and distance from us is constant. These discoveries became Hubble's Law.

In music, the basic idea of Schoenberg's twelve-tone row would grow to become music's metaphor to the unified field theory of Einstein. By serializing the idea of number and order to every proportion of time, pitch, intensity, and duration, composers of the 1950s and 60s created a musical universe that had its corollary in Hubble's theory of the shape of the universe.

In one of the most comprehensive statements of the serial musical universe, Ernst Krenek, who was living in Tujunga not far from Mount Wilson, adopted the sestina, a poetic form found in Dante and Petrarch, to create an audible realization of the possibilities facing music.

> *Bygone are sound and mourning, tender stream.*
> *Vibrations of the second becomes measure.*
> *What lives in history, was it only chance?*
> *Decline, fading sound, vanished shape?*
> *The hour causes change, turns the time.*
> *What looks ahead subordinates itself to number.*

Krenek's *Sestina* consists of six stanzas highlighting the words that appear at the end of each line. If in the order of the first stanza the words are 1-2-3-4-5-6, the second rotation will become 6-1-5-2-4-3. This structure will proceed until the cycle is complete. The six words emphasize shared characteristics of music and mathematics.

Stream

Measure

Chance

Shape

Time

Number

7. Ernst Krenek,
    Sestina (1957),
    Bärenreiter,
    Kassel, Germany.

The elements of pitch, duration, density, spacing, speed, and dynamics become the unified field of the composition. Krenek himself would point out that "the paradox of ultimate necessity's causing unpredictable chance is the topic of the *Sestina*."[7]

The other important observation to come from Hubble's Law is that the universe is finite. Nothing is more than twenty billion light-years away. Though twenty billion light-years are a difficult time frame to comprehend in the real world, it probably is no more audacious to us than the computations of Kircher were to the 17th century. From this *Ars Combinatoria* of music and mathematics emerged both a paradoxical finite and infinite articulation of the possibilities of the worlds of heaven and earth.

The Vector of Ignorance: Musica Instrumentalis

The only vestige of the educational traditions of the quadrivium active today is Musica Instrumentalis. But as the concepts of science and music have become more expansive, and in the short run complicated, there is a tendency for both disciplines to pass in the night. Music recedes more and more into sentimental convention, and science seems only to career forward. Both disciplines, in diametrically opposite ways, seem impatient with their history and background. These ironies are what I call the vector of ignorance.

We have experienced first hand the failure of faster, better, and cheaper. In discussions I have had with prominent figures from the world of science, the common situation we have defined between the arts and sciences is their tendency to absent intellectual rigor from their study and contemplation. When intellectual rigor is missing, these disciplines devolve into a prosecution of craft alone. No matter how brilliant the results of craftmakers, there is the real danger of creating a counterfeit majority.

The other very real situation facing both the arts and sciences is the fact the concepts we are dealing with at the beginning of the Third Millennium are hard for the average person to get their arms around. The accumulation of knowledge is formidable. This leaves our educational systems confused as to the best way to proceed with the integration of past and present. Here is where the role of history can come onto center stage. History can and should function as a discussion about the future that takes place in the past.

Dennis Richard Danielson, in his *Book of the Cosmos*, writes "One anxiety we understandably feel in reading about "old" ideas is that they may be wrong, outdated, superseded. Leaving aside the neglected truism that today's up-to-date ideas may appear wrong, outdated, and superseded ten years from now, I think there are ways of approaching conceptions from the past without condescension and at the same time without disregard for the question of truth."[8]

8. Richard Dennis Danielson, *The Book of the Cosmos* (Perseus Publishing, 2000).

As music seems to lose the scientific foundation that it could take for granted in past eras, science runs the risk of marginalizing the mystery of its heritage. When this is translated into the educational arena, the results are that music remains in the past and science projects only into the future. If we are to move away from faster, better, and cheaper, this vector of ignorance needs to change.

If one looks at the historical imperative of the developments that have brought the arts and sciences to their present degree of achievement, one will begin to hope that the best thing we can do is regenerate intellectual intuition. Perhaps things truly are connected.

It is abundantly clear that the study and knowledge of one discipline brings clarity and comprehension to the other. Just when one seems overwhelmed by the concepts of Riemann and Einstein, a fresh perspective on Beethoven brings a much-needed focus. When layer upon layer of sound and event in 20th century music becomes dizzying, the beauty of Hubble's law provides insight and foundation.

We still have much to learn from the traditions of the past, for if nothing else they provide a window in time to the ideas and concepts of our own era and future. As we begin the Third Millennium and look out over the great expanse of human experience, we would do well to instigate a complete reconciliation between the arts and sciences. Together, they allow us to "Explore with Desire" the expanses of the universe and our earthly existence in the cosmos.

Jeff von der Schmidt is artistic director of Southwest Chamber Music. "The Melody of Numbers" was previously published in the Southwest Chamber Music program book "The Universe: Music of the Spheres."

Creation
Constellations
and the Cosmos

This collective exhibition on the theme of the universe offers a rare opportunity for the Norton Simon Museum to bring together diverse aspects of its vast holdings. For the exhibition *Creation, Constellations, and the Cosmos,* we have narrowed down this broad topic to focus on four theological constructs. The first section, "Axis Mundi," addresses artistic representations of the cosmic axis that serves as the symbolic connection between heaven and earth. The second section, "Light Symbolism," explores halos, the markers of divine status that adorn Christian as well as Buddhist and Hindu gods. The third section, "Cosmic Circles," examines how this perfect shape takes on a sacred quality. Lastly, "Constellations" focuses on artistic depictions of solar and lunar motifs.

Axis Mundi

The concept of the cosmic axis is part of a complex cosmological and mythological tradition. The *axis mundi* symbolizes the separation between earth and the heavens as well as the connection between the two realms. It can take the form of a tree, pillar or pole, ladder, lingam or phallus, celestial ray, wand or staff, or mountain. The role of this symbol in diverse religions and cultures has been described by scholars of comparative mythology, most notably Mircea Eliade: "The symbolism of the axis mundi is complex: the axis supports the sky and is also the means of communication between heaven and earth. When he is close to an axis mundi, which is regarded as the center of the world, man can communicate with the heavenly powers."[1]

A familiar story from Genesis is that of Jacob's Ladder. Jacob lay down upon the rocky ground to sleep and dreamed of a ladder that was set upon the earth and reached up to heaven. He beheld angels of God ascending and descending the

Fig. 19
Descent of the Buddha (detail) Eastern Tibet (Kham), 19th century Pigments and gold on cotton; 18 x 12 1/4 in. Norton Simon Museum, Pasadena, California, Gift of Dr. Pratap and Chitra Pal, 1998

1. Mircea Eliade, *Symbolism, the Sacred, and the Arts,* ed. Diane Apostolos-Cappadona (New York: Crossroad, 1988), 99.

ladder, at the top of which stood God, who spoke to him. The ladder also appears
prominently in the life of Buddha Shakyamuni, the founder of the Buddhist religion.
Queen Maya, the Buddha's mother, died when he was seven days old and therefore
never received her son's teachings. After attaining Enlightenment, the Buddha preached
to his mother in heaven, after which he commanded that a triple ladder be built so
that he could descend bodily from heaven to earth to continue his ministration.
One of twelve great deeds of the Buddha, his descent is depicted in a well-preserved

Fig. 20
Buddhist Triad with
Shakyamuni
India, Bihar (Gaya
region), c. 1000
Schist;
26 ¾ x 17 ⅛ in.
The Norton Simon
Foundation,
Pasadena, California

nineteenth-century painting from eastern Tibet (fig. 19). In the upper right corner, the celestial realm is depicted as a heavenly paradise, and the Buddha is seated in a shrine or pavilion preaching to a small gathering. Below on earth, additional figures bearing gifts await his arrival.

Another *axis mundi* from the Buddhist tradition is the bodhi tree, under which Siddhartha attained Enlightenment and became the Buddha. As Eliade reminds us: "The temple, the Cosmic Mountain, the Pillar, the Tree—all these symbols are equivalent. They support the world, they are the Axis of the Universe, they are the center of the world."[2] A carved stone relief from eastern India (c. 1000; fig. 20) depicts the Buddha seated beneath a bodhi tree, accompanied by two bodhisattvas. The tree is represented by two stylized leaves that flank his *ushnisha*, or cranial protuberance of wisdom. Above his head is a parasol, which is another symbol of the Buddha's transcendental nature.

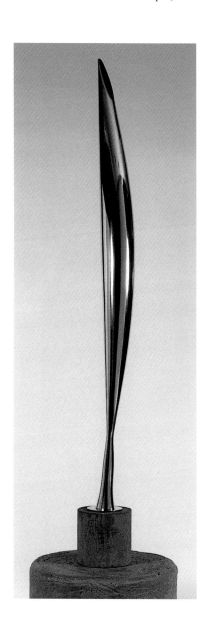

The modernist sculptor Constantin Brancusi was influenced by folk mythology. *Bird in Space* (1931; fig. 21) is one of a series of bird sculptures that he worked on over the course of twenty-eight years. Brancusi was interested in the theme of ascension and believed his streamlined bird in magical flight symbolized freedom and bliss. The verticality of the bird's flight suggests a gleaming *axis mundi*. As Eliade has noted, "The symbolism of flight expresses an escape from the universe of everyday experience, and the double intentionality of that escape is obvious: it is at the same time transcendence and freedom that one obtains by 'flight.'"[3]

2. Ibid., 137.

Fig. 21
Constantin Brancusi
(1876–1957)
Bird in Space, 1931
Polished bronze
with cylindrical
limestone base;
height: 73 in.
The Norton Simon
Foundation,
Pasadena, California

3. Ibid., 101.

The Halo:
A Symbol of Light

Halos are common symbols in both Christian and Asian religious imagery. The halo is emblematic of the sun, its round shape and golden color symbolizing divine radiance. Gerard David's *Coronation of the Virgin* (c. 1515;

Fig. 22
Gerard David
(c. 1450–1523)
*The Coronation of
the Virgin*, c. 1515
Oil on panel;
27 7/8 x 21 1/4 in.
The Norton Simon
Foundation,
Pasadena, California

Fig. 23
*The Future Buddha
Maitreya Flanked by
the Eighth Dalai
Lama and His Tutor*
Tibet, 1793–94
Silk appliqué;
268 x 177 in.
The Norton Simon
Foundation,
Pasadena, California

fig. 22) is a formal portrait of Mary with the infant Jesus, accompanied by four prophets and two angels. The splendid gold background of this painting is in fact a large halo that refers to the radiance of the sun. Buddhist gods are also given halos, as seen in the monumental appliqué *The Future Buddha Maitreya* (fig. 23). Maitreya not only bears a halo around his head but displays a larger body halo as well. In both the Christian and Buddhist traditions, the golden color represents illumination and sacredness. The great Hindu god Shiva, when depicted as Lord of the Dance (Nataraja), is surrounded by a halo of flames, which represents the cosmos (see fig. 24).

Fig. 24
Shiva as Lord of the Dance (Nataraja)
India, Tamil Nadu,
c. 1200
Bronze;
29 1/4 x 23 1/2 in.
The Norton Simon
Foundation,
Pasadena, California

Cosmic Circles

The circle is a symbol that transcends cultures, religions, and time.
Over the ages the circle has appeared as a cosmic emblem in spiritual imagery, as well
as a dominant geometric form in abstract art. Among the most important and familiar
icons of the Hindu religion is the god Shiva as Nataraja, or Lord of the Dance.
Nataraja creates the cosmos with the "Dance of Furious Bliss" *(ananda-tandava)*,
which is performed in the center of a circle symbolic of the universe. The all-powerful god dances to the sound of the cosmos's heartbeat *(maya)*, produced by the hourglass-shaped drum held in his upper right hand. Shiva's creative powers are balanced by his ability to destroy. The deity's simultaneous destruction of the universe is signified by the flames lining the outer edge of the circle *(tiruvasi)* and by the single flame held in his left hand.

Fig. 25
Wassily Kandinsky
(1866–1944)
Heavy Circles, 1927
Oil on canvas;
22 ½ x 20 ½ in.
Norton Simon
Museum, Pasadena,
California, The Blue
Four Galka Scheyer
Collection, 1953

The circular forms in the abstract paintings of Wassily Kandinsky were also intended as symbolic links to the cosmos. The circle, a key component of Kandinsky's pictorial vocabulary, was used to create visual harmonies expressive of the artist's feeling for the mystical "music of the spheres."[4] He viewed visual rhythm as a fundamental principle in art, which linked the visible with the invisible and united a work of art with the cosmic realm.[5] The artist explained that while the circle was an important formal element in his work, "the reason . . . is not the 'geometrical' form . . . but rather my own extreme sensitivity to the inner force of the circle in all its countless variations."[6] The circle appears as the sole motif in Kandinsky's *Heavy Circles* (1927; fig. 25). In this painting, fifteen

4. Vivian Barnett, curatorial files, Norton Simon Museum.

5. Dee Reynolds, "Symbolist Aesthetics and Early Abstract Art: Sites of Imaginary Space" (New York: Cambridge University Press, 1995), 199.

6. Kenneth C. Lindsay and Peter Vergo, eds., "Kandinsky: Complete Writings on Art" (Boston: G. K. Hall, 1982), 740.

brightly colored circles, in various sizes and hues, are displayed against a dark background in an arrangement that strongly resembles a planetary rotation. The two largest and most central circles, suggestive of the sun and moon, form an eclipse, while the surrounding circles suggest planets.

Robert Irwin's circular disc "paintings" of the 1960s also evoke a cosmic sensibility. Made of cast acrylic, these illusory discs seem to hover on the wall without visible support. Direct illumination by spotlight makes the physical boundaries of the discs indiscernible from the shadows cast upon the wall behind, suggesting an infinite expanse of light and space. Although Irwin's primary concerns in the disc paintings were with perception and how the object relates to the space it occupies, his ethereal paintings appear more celestial than terrestrial.[7]

7. Lawrence Weschler, *Seeing Is Forgetting the Name of the Thing One Sees: A Life of the Contemporary Artist Robert Irwin* (Berkeley: University of California Press, 1982), 98–109.

Constellations

In humankind's attempt to better understand the universe and our place within it, we have often looked to the stars. In Christian iconography, the sun is symbolic of the Heavenly Father, while the coupling of the sun and moon refer to the Virgin Mary. In the twelfth chapter of the Bible's book of Revelation, Saint John envisions "a woman adorned with the sun, standing on the moon with twelve stars on her head for a crown." In Gerard David's *Coronation of the Virgin*, Mary appears as Saint John's "Woman of the Apocalypse." The lavish gold background refers to the radiance of the sun and God's divine presence. The crown held above Mary's head by two angels signifies her position as Queen of Heaven. The crescent moon beneath her feet symbolizes chastity.

In Eastern religious traditions, constellations are often associated with divine forces. Celestial figures such as those seen on the early tenth-century Cambodian sculpture *Stele with Five Planetary Deities* (fig. 26) play an important role in the religious teachings of both Hinduism and Buddhism. This stele fragment is believed to have originally shown nine deities *(navagraha)* and would have embellished a Cambodian temple.[8] The five planetary gods represented include Surya (sun), Soma (moon), Mangala (Mars), Budha (Mercury), and Brihaspati (Jupiter). As is customary in Khmer depictions, these celestial figures are identical and appear mounted on the backs of animals in conventional poses.

8. Pratapaditya Pal, curatorial files, Norton Simon Museum.

Fig. 26
Stele with Five Planetary Deities
Cambodia, Angkor Period, early 10th century
Sandstone;
17 1/2 x 49 in.
The Norton Simon Foundation, Pasadena, California

Celebrated American sculptor Alexander Calder took inspiration from the cosmos, stating that "the underlying sense of form in my work has been the system of the Universe, or part thereof."[9] This lifelong interest was manifested in his art through shapes representing the sun, moon, planets, and stars. While Calder is best known as the inventor of the mobile in the 1930s, he also produced stunning gouaches of celestial forms (see fig. 27).

Well before humankind's giant step to the moon in 1969, June Wayne was exploring outer space in her paintings, graphics, and textiles. Wayne's fascination with the universe was fueled by the exciting scientific discoveries of the last half of the twentieth century, which led to her artistic investigations of such concepts as time, genetic coding, molecular energy, the solar system, and the individual's place within the complex structure of the universe. Her 1965 lithograph *At Last a Thousand I* (fig. 28) anticipates the photographs later taken of the Andromeda Galaxy by NASA. This work is suggestive of an atomic explosion, as well as the shifting structures of galaxies and planetary systems.[10] According to Wayne, the twinkling lights visible throughout the composition refer to lemmings, known for their mass migrations and suicides.[11] Here the lemmings move in a circular pattern, awaiting an unknown fate, some having already fallen into the void of deep space.

9. Jean Lipman, *Calder's Universe* (New York: Viking Press; Whitney Museum of American Art, 1976), 18.

10. Arlene Raven, "June Wayne: Tunnel of the Senses," in *June Wayne: A Retrospective* (Purchase: Neuberger Museum of Art, Purchase College, State University of New York, 1997), 44, 55.

11. June Wayne, conversation with Michelle Deziel at the artist's studio, Los Angeles, 1 September 2000.

Fig. 27
Alexander Calder
(1898–1976)
Maelstrom with Blue,
1967
Gouache on paper;
43 x 29 ¼ in.
Norton Simon
Museum, Pasadena,
California, Gift of
Mr. W. H. Hal
Hinkle, New York,
1986

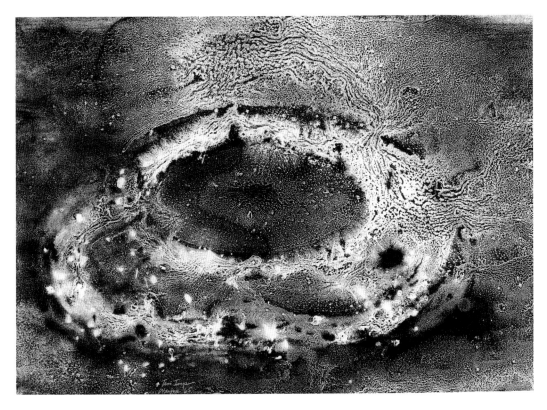

Fig. 28
June C. Wayne
(b. 1918)
At Last a Thousand I,
1965
Lithograph;
24 x 34 in.
Norton Simon
Museum,
Pasadena, California,
Anonymous Gift,
1969

These examples from the collection of the Norton Simon Museum
suggest the universality of cosmic symbolism and its ability to take on many different
meanings and encapsulate diverse worldviews, across cultures and through the ages.

Michelle Deziel is curatorial assistant, and Christine Knoke is assistant curator
of Asian art at the Norton Simon Museum, Pasadena, California.

Constructing *the* Cosmos *in the* Religious Arts *of* Asia

Gaze then, thou Son of Pritha!
　　　　I manifest for thee
　　　　　　　Those hundred thousand thousand shapes
　　　　that clothe My mystery.
　　　Krishna to Arjuna, from the *Bhagavad Gita*[1]

P acific Asia Museum's exhibition *Constructing the Cosmos in the Religious Arts of Asia* looks at some of the ways in which four ancient religions born in Asia—Daoism, Hinduism, Jainism, and Buddhism—have perceived the universe and depicted it in their art.[2] It features models, diagrams, and symbols of the cosmos created over the centuries by religious artists. A contemporary three-dimensional Tibetan Buddhist mandala has also been included to highlight the importance of the process of construction of such cosmic models and to demonstrate the universal and timeless nature of the quest for spiritual enlightenment.

　　　　　It is impossible to do justice here to the feast of cosmological beliefs and related imagery of these four great religions. The exhibition and this essay serve more as an hors d'oeuvre, a sampling of some of the exquisite yet powerful cosmic imagery and symbolism that Asian religions have inspired.

Daoism

　　　　　Daoism (or Taoism), one of the two major religious and philosophical systems developed in China, is said to have been founded by the philosopher Laozi (Lao-tzu) around the sixth century B.C. but is probably considerably older. Its main text, the poetic *Daode jing (Tao-te-Ching)*, describes the power of the Dao (or Tao),

1. *Bhagavad Gita*, chap. 9; translated by Sir Edwin Arnold; the *Bhagavad Gita* (The song celestial) is an episode from the great Hindu epic, *Mahabharata*, written in Sanskrit more than two thousand years ago.

2. Since an exhibition of the art of all of the religions of Asia—including Islam, Christianity, and smaller native religions—would be a feat of truly cosmic proportions, the scope of this exhibition has been necessarily limited to these four Asian-born religions.

Fig. 29
Krishna in His Cosmic Form (Vishvarupa)
India, 19th century
Ink and colors on paper; 52 x 35 ½ in.
Collection Julia McDivitt Emerson

Fig. 30 (above)
Porcelain Bowl
China, Qing dynasty
(1644–1911)
Blue and white
porcelain with
celadon glaze;
diameter: 2 3/4 in.
Pacific Asia Museum
Collection, Gift of
Robert and Sheila
Snukal, 1996

Fig. 31 (right)
Mountain Landscape
China, Ming dynasty,
1637
Ink on silk;
63 x 22 in.
Pacific Asia Museum
Collection, purchased
with funds provided
by Mr. and Mrs.
Alan Braun, 1987

3. John Blofeld,
*Taoism: The Quest
for Immortality*
(London: Unwin
Hyman, 1979), 2.

the "way," the universal principle that engenders all things and is at the same time present in all things. Originally largely philosophical, Daoist teachings stressed contemplation, mystical union with nature, and the concept of letting things take their natural course.

Around the third century B.C., Daoism developed a more religious aspect, and beliefs about the afterlife and the soul evolved. Daoists believed that the soul split into two after death; one part, the *po*, lingering with the corpse, while the *hun* journeyed in search of Paradise. In the Daoist cosmology there was also an Underworld, the Yellow Springs, where the *hun* might also stray. Later many folk deities were gradually incorporated into Daoist beliefs. Still practiced in China, Daoism has also permeated Confucianism and Buddhism, in particular Zen Buddhism.

According to the *Daode jing*, the Dao is the universal force of the cosmos. This force operates through the continuous interplay of two opposing forces or energies, yin and yang—the former negative, passive, and female; the latter, positive, active, and male. These two forces are symbolized by a circle (the Dao) formed by two intertwined comma shapes, one black, the other white and each containing the seed or essence of the other. The well-known yin-yang symbol graces many Daoist texts and objects and many of the decorative arts of China (see fig. 30).

Perceived as "unknowable, vast, eternal," and the very goal of existence itself,[3] the Dao is too abstract for words or images. Yet the presence of the Dao, and of yin and yang, can be sensed in many Chinese monochrome landscape paintings (see fig. 31). In Daoist thought, pure, or cosmic, yang pertains to heaven, while pure, cosmic yin pertains to the earth, and the two must connect. For centuries Daoists have believed that at certain places on earth invisible lines, or "dragon veins," run down from the sky into the mountains and along the earth, channeling cosmic yang energy toward the cosmic yin energy. The bold, dark brushstrokes delineating the mountains and rivers of Chinese landscape paintings recall these veins of cosmic energy. Tiny human figures and their mountain abodes, dwarfed by their natural surroundings, also remind the viewer of the immensity of the Dao.

Fig. 32
Dragon Robe
China, Qing dynasty
(1644–1911)
Embroidered silk;
height: 58 ½ in.
Pacific Asia Museum
Collection, Gift of
Mr. and Mrs. M.
Curtis Smith, 1976

In Daoist symbolism, the dragon, an inhabitant of both sky and sea, was credited with rainmaking and other cosmic powers. As such, it represented heavenly cosmic yang and was closely associated with the Chinese emperor, the Son of Heaven. During the Qing dynasty (1644–1911), Chinese emperors wore

Fig. 33
The Churning of the Ocean of Milk
Panjab Hills, Kangra, India, c. 1785
Ink and gouache on paper;
10 1/2 x 7 1/16 in.
San Diego Museum of Art, Edwin Binney 3rd Collection, acquired 1990

elaborate silk robes embroidered with dragons as an emblem of their own cosmic power (see fig. 32). Embroidered with mountains rising out of the sea and cranes flying among clouds, the robes represented the terrestrial realm. When the emperor, who represented heaven, wore the dragon robes, he symbolically united heaven and earth.

Hinduism

In India the religion known as Hinduism has developed over thousands of years into a complex religious and social tradition. Fundamental to Hinduism is the idea of *moksha*, the release of the soul from the perpetual cycle of rebirth, known as *samsara*. In Hindu teachings, all beings are chained to this cycle, and one's actions, or karma, determine the level of one's rebirth. Good deeds in one's present life lead to a higher rebirth; immoral behavior, a lower rebirth. Similarly, one's present circumstances are the result of actions in a previous existence and are almost impossible to alter.

Early Hindu ideas about the cosmos are found in the sacred texts known as the Vedas,[4] some containing ideas that connect humanity with Brahman, the underlying universal principle, much like the Chinese Dao. According to these texts,

4. A collection of sacred hymns and philosophical texts based on older oral teachings of the early Aryan settlers in India.

this "impersonal, and ultimately unknowable, controlling force" exists beyond even the gods and determines the entire cosmic order.[5] Priests known as *brahmana*, or brahmins, performed sacrifices to Brahman, which symbolized the act of universal creation. As intermediaries between men and this cosmic force, the brahmins attained the highest position in the rigid social order.

5. Richard Blurton, *Hindu Art* (London: British Museum, 1992), 28.

As the worship of gods and goddesses became more predominant in Hindu practice, the concept of the cosmic Brahman came to be represented by the anthropomorphic creator god Brahma, who often appears in images of the creation of the cosmos. In one, Brahma and other gods and demons create the world by churning up the milky cosmic ocean, just as humans churn milk to produce butter (fig. 33). The gods and demons pull on the churning rope (actually a snake), producing various gods and goddess, magical liquids, and objects from the depths of the cosmic ocean.

Perhaps the most powerful cosmic image in Hinduism is *Shiva Nataraja* (*Shiva, Lord of the Dance*), in which Shiva, lord of creation and destruction, dances the cosmic dance (see fig. 34). With one hand holding the fire of destruction and another wielding the double-sided drum of creation, Shiva summons the end of one cosmic cycle and the beginning of another in one ecstatic dance. In this form, he also symbolizes the dual forces at work in the universe: creation and destruction, good and evil, day and night, male and female, and love and hate.

Fig. 34
Shiva as Lord of the Dance (Nataraja)
India, c. 1200
Bronze;
29 1/4 x 23 x 9 in.
Norton Simon
Museum, Pasadena,
California

Another major Hindu god, Vishnu, preserves and maintains cosmic order. In the *Bhagavad Gita*, Vishnu, in the form of the god Krishna, teaches the warrior prince Arjuna the importance of dharma, a crucial Hindu concept that encompasses both social duty and the law of the universe. To illustrate this law,

Krishna astounds Arjuna by revealing the entire universe—the heavens above, the human realm in the middle, and the hells below—inside his divine body (see fig. 1).

Jainism

Around the sixth century B.C. in India, many spiritual and philosophical leaders rejected the teachings of early Hinduism. One such leader was Mahavira, the founder of Jainism. Jains share the Hindu belief in the cycle of rebirth and karma, believing that eventual release from this cycle is possible by perfecting one's soul through successive lives. Those beings who have perfected their souls and transcended the world of all beings, including gods, are known as Jinas (saints) or *tirthankaras* (crossing-builders). There have been twenty-four Jinas in our current age, the most recent of whom was Mahavira. Many Jains practice strict asceticism and *ahimsa* (nonviolence) toward all living creatures with the aim of perfecting their souls.

One of the most intriguing representations of the concepts of karma and the cycle of rebirth is the game of snakes-and-ladders. Played in India, Nepal, and Tibet by Hindus, Jains, and, more rarely, Muslims, the game consists of a checkered board, made up of eighty-four squares in Jain examples, known as *gyanbazi*, or the game of knowledge (see fig. 35).[6] The squares represent progress through life and contain words about moral conduct. Stretching across these squares are snakes and ladders, denoting bad and good conduct, respectively. Players throw dice to progress toward the heavens, shown as a pavilion at the top of the board. Bad conduct is rewarded by slow progress, and good conduct has the opposite result, simulating the concept of karma. Traditionally a didactic game, it evolved into a recreational pursuit, eventually fascinating British colonials so much that they took the game home with them. In the United States, it evolved into the less frightening chutes-and-ladders.

Many cosmological texts and diagrams were created to help Jain worshipers attain spiritual liberation. Though often elaborate, Jain diagrams of the world simplify complex teachings about the transmigration of the soul.[7] According to Jain cosmology, the universe is divided into three realms: the upper realm of the gods, the lower realm of the damned, and the middle realm, where humans can be born and reborn. These three realms can be seen in Jain images of the cosmic man, or Lokapurusha, in which the cosmos is superimposed on a human body. At the figure's waist is the middle realm of mortals.

6. See Deepak Shimkhada, "A Preliminary Study of the Game of Karma in India, Nepal, and Tibet," *Artibus Asiae* 44 (1983): 4

7. For more details, see Collette Caillat and Ravi Kumar, *The Jain Cosmology* (Basel: Harmony Books, 1981).

Fig. 35
Snakes and Ladders (Gyanbazi)
India, 19th century
Ink and colors on paper; 23 x 19 3/8 in.
Collection of Julia McDivitt Emerson

Fig. 36
Jain Diagram of the Realm of the Mortals
India, 18th–19th century
Ink and colors on cloth; 36 x 36 in.
Collection of Sam Fogg, London

This middle realm, or Manushyaloka (the world of men), is the focus of many elaborate Jain diagrams (see fig. 36). The realm consists of two and a half continents. In its very center is the cosmic mountain of Jainism and Buddhism, Meru, located on Jambudvipa (the "Continent of the Rose-Apple Tree"). All around Mount Meru, represented as a circle in the center, are parallel bands representing mountain ranges dividing the continent into separate countries, each containing rivers. Encircling Jambudvipa is Lavanasamudra, the "Sea of Salt," shown as a blue ring containing fish and other aquatic creatures.

Buddhism

India of the sixth century B.C. was also the setting for the life of Siddhartha Gautama, or the Buddha ("the Enlightened One"), the founder of Buddhism. The Buddha taught that release from the cycle of rebirth can be achieved by understanding that in all life there is suffering, which is caused by desire and attachment. To end suffering, one must transcend desire and attachment by practicing a life of moral behavior and meditation. Gradually Buddhism divided into three distinct traditions, or vehicles (*yana*): the first, a monastic tradition focusing on the pursuit of individual enlightenment; the second, the Mahayana ("greater vehicle") tradition, emphasizing compassion and enlightenment for all beings; the third, Vajrayana ("diamond/thunderbolt vehicle") Buddhism, another altruistic tradition, but one that employs secret, mystical practices to transform human passions into a path toward enlightenment.

Fig. 37
Stupa
Nepal, c. 1600
Gilt copper, silver;
height: 6 ¾ in.
Pacific Asia Museum
Collection, Gift of
David Kamansky in
memory of Carolyn
Houghton, 1997

One of the most potent symbols of Buddhism and the Buddhist cosmos is the stupa, a monument built to house the relics of a deceased buddha, or a Buddhist teacher (see fig. 37). For more than two thousand years, Buddhists have erected large and small stupas—pagodas in East Asia—as reminders of death and thus the impermanence of existence and the goal of enlightenment. To many, its whole form represents the cosmos itself, or the cosmic Mount Meru, while its parts symbolize the six elements of the Buddhist universe: the square base, earth; the rounded dome, water; the rising spire, fire; the umbrellas, wind; the spire at the top, space. The whole also represents the sixth element, the buddha-mind.[8]

8. Denise Patry Leidy and Robert A. F. Thurman, *Mandala: The Architecture of Enlightenment* (New York: Asia Society Galleries, Tibet House, Shambhala, 1997), 50.

9. Elizabeth ten
Grotenhuis, *Japanese
Mandalas:
Representations of
Sacred Geography*
(Honolulu: University
of Hawai'i Press,
1999).

10. Ibid., 127.

11. Leidy and Thurman,
Mandala, 168.

Related to the idea of the stupa is the Buddhist mandala. The Sanskrit word *mandala*, also used in Hinduism and Jainism, literally means a sacred center (*la*) that has been set apart or adorned (*mand*).[9] In theory, therefore, a mandala can take any form. A mandala is usually a painted, printed, or embroidered geometrical pattern, however, often resembling the floor plan of a palace, with central and outer halls and four main gates.

For Mahayana and Vajrayana Buddhists, a mandala is a model of a perfected universe, a "blueprint for buddhahood,"[10] or enlightenment, a sort of cosmic map of the sacred abode of deities. This Tibetan mandala (fig. 38) is the abode of Yamantaka, a wrathful, buffalo-headed protector deity of the Tibetan Buddhist pantheon. A practitioner of Tibetan Vajrayana Buddhism, attempting to attain the type of enlightenment associated with this deity, strives to visualize every detail of the architecture and environment of the mandala and mentally transform the two-dimensional mandala into the three-dimensional realm of the deity.[11]

Three-dimensional mandalas also exist but are rarer. In Tibet, most large Buddhist monasteries housed three-dimensional mandalas, but few remain today. In 2000, however, a Tibetan Buddhist community based in Los Angeles and led by Lama Chodak Gyatso Nubpa embarked on a project to build a three-dimensional mandala (fig. 39), with the dual aims of preserving a threatened tradition and of employing the mandala as a tool for universal peace. According to Lama Gyatso, this mandala, a type known as the Shi-tro Mandala, is the celestial realm of one hundred Buddha Families and represents the enlightened qualities and compassion already within us. Simply by looking at it, the viewer can sow the seeds to attain ultimate spiritual release.

Fig. 38
Yamantaka Mandala
(detail)
Eastern Tibet,
c. 1700
Ink and colors on
silk; 34 x 23 in.
Pacific Asia Museum
Collection,
Gift of Dr. and Mrs.
Jesse Greenstein,
1996

Fig. 39
Shi-tro Mandala with
Lama Chodak Gyatso
Nubpa (left) and
artist Pema Namdol
Thaye (right),
Los Angeles, 2000
Painted wood
and plaster;
10 x 10 x 10 in.
(approx.)
On loan from the
Chagdud Gonpa
T'hondup Ling,
Los Angeles

Precise knowledge of the structure of mandalas, as well as remarkable artistic skill, is required to build such an elaborate object. In addition to being a sacred work of art, this mandala is also a work of great devotion and respect. During the nine-month period of construction, Pema Namdol Thaye, one of Tibet's most talented Buddhist artists, and his wife and brother, meditated each morning to attain the pure mental state necessary to construct this mansion for the deities.[12]

12. Teresa Watanabe, "Prophecy, Karma, and a Buddhist Icon in Glendale," *Los Angeles Times*, 26 June 2000.

In this work, as in most of the works in the exhibition, the aim of the artist was not so much to produce an artistic masterpiece as to create an image or model of the cosmos to enable viewers to better understand the world and themselves. Looking at these cosmic images, devotees discover an aspect of themselves. These diverse creations of wood, paper, metal, silk, and paint from the Daoist, Hindu, Jain, and Buddhist traditions, more than explaining the true, physical nature of the universe, help the individual devotee find his or her own place in the mysterious vastness of the cosmos.

Meher McArthur is curator of Asian art at the Pacifica Asia Museum.

George Ellery Hale *and the* Development *of* Southern California

Geore Ellery Hale (1868–1938) was a complex, intense, and extremely driven man who straddled several important worlds simultaneously. On the one hand, he played a vital role in the development of twentieth-century astronomy, and on the other, he served as a driving force in the areas of culture and education in Southern California. As an astronomer, he was responsible for worldview-changing discoveries in solar physics. But he was also involved with the development—as an administrator, fund-raiser, and friend of the rich and famous—of three of the most important observatories of the past century, each of which had the largest telescope in the world for a considerable period. He served as director of the Yerkes Observatory in Chicago from 1897 to 1903. Next he started the Mount Wilson Observatory, northeast of Pasadena, with a grant of $10,000 from his friend Andrew Carnegie, running it from 1904 until his death in 1938. In his final years he also spearheaded the effort to establish an observatory on Mount Palomar, near San Diego. The Palomar Observatory would become the home of the world's largest telescope, the two-hundred-inch Hale Telescope (named in his honor a decade after his death).

Hale, the statesman of science, would have been pleased. In his spare time he founded the important *Astrophysical Journal*, now in its 102nd year. He also invented the spectroheliograph, an instrument that allowed unparalleled analysis of the structure and activity of the surface of the Sun. The spectroheliograph represented perhaps the most important advance in solar astronomy since the time of Galileo.

Fig. 41 (below) Andrew Carnegie (left) and George Ellery Hale outside the sixty-inch telescope dome on Mount Wilson, March 1910 Collection The Henry E. Huntington Library and Art Gallery, San Marino, California

Fig. 40 (opposite) Left to right: A. A. Noyes, Hale, and Robert Millikan (the first important president of the California Institute of Technology) on the terrace of the Gates Chemical Lab at Caltech. Collection The Henry E. Huntington Library and Art Gallery, San Marino, California.

Fig. 42
Hale seated at his desk, probably at his office
(room #1) at the original Mount Wilson
"Monastery" building, c. 1905
Collection The Henry E. Huntington Library and
Art Gallery, San Marino, California

Fig. 43
Carnegie (standing) and Hale (seated to
Carnegie's right) at banquet, March 1910
Collection The Henry E. Huntington Library and
Art Gallery, San Marino, California

1. Mary Browne Hale to her mother,
   Mrs. Philemon Scranton, 12 October 1883,
   cited in Helen Wright, *Explorer of the
   Universe: A Biography of George Ellery Hale*
   (New York: Dutton, 1966), 31.

But Hale found deep interest and motivation
in other arenas besides astronomy. He was a close friend
of Henry E. Huntington and one of the first trustees of the
Huntington Library. He deserves at least partial credit for
convincing Huntington to leave his extraordinary estate as
a public space for scholars and visitors. He was also
instrumental in turning the small, provincial Throop Polytechnic Institute in Pasadena into the California Institute of
Technology (Caltech), now one of the country's finest
universities and producer of numerous Nobel laureates.
This essay focuses primarily on those two contributions.

Hale's social circle was an important
one. He was a friend of Edwin Hubble, the great astronomer. Hale, older than Hubble by some twenty-one years,
was already the director of the Mount Wilson Observatory
when young Hubble was completing his doctoral disser-
tation at the University of Chicago and about to enter
World War I as a soldier in 1917. Hale promised Hubble
that he would have a place at Mount Wilson after
returning from the army, thus setting the course for some
of the century's most important astronomical discoveries.

Born in Chicago just after the Civil War,
Hale developed a precocious interest in scientific pursuits.
While other children played, he looked through
microscopes, and his little bedroom was soon overflowing
with books, botanical specimens, and tools. He finally
convinced his mother to allow him to turn an upstairs
room into a laboratory. "George spends all the time he can
get in his shop," his mother noted about her fifteen-year-
old son. "We hope he can learn a great deal from his
'tinkering.'"[1] His father purchased George's first telescope
in 1882, just in time for him to watch the Transit of
Venus, in which the planet crossed in front of the Sun.

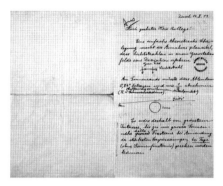

Hale was hooked, and his interest in things astronomical never wavered. He quickly worked his way through the Massachusetts Institute of Technology. It was in Southern California that he found his most rewarding work and a life of great satisfaction, undertaken with the intense energy that was his hallmark. He was a scientist first, but one with a profound interest in the humanities. Hale was a benevolent opportunist, and just as he had attached himself to Andrew Carnegie a decade or so earlier, he became friends with Huntington, after "finding himself" seated next to him at the head table at the Hotel Maryland in Pasadena in 1906. Hale discussed the potential benefits of establishing a research library based on Huntington's book and manuscript collecting interests. Huntington listened but was unmoved, and it was not until 1914 that the two men became better acquainted. In mid-April of that year Hale wrote to Huntington's new bride, Arabella, asking about her already-successful son Archer, who was wealthy and becoming known as a man with philanthropic instincts. His approach to the Huntingtons through Arabella worked, and eventually Archer and Hale met in New York to discuss the possibility of promoting culture and art— issues always foremost in Hale's mind. Coincidentally, or perhaps not so coincidentally, Henry Huntington was in town that day as well, and Hale dined with him. Hale again broached the idea of the research library. Mr. Huntington stated that he planned to leave his San Marino estate and collections to the board of supervisors of Los Angeles County. Hale was aghast but said nothing. The next day, however, he wrote Huntington a long letter warning him about the dangers of relinquishing control of the

Fig. 44
Letter from Albert Einstein to Hale, 1913. Einstein's general theory of relativity implied that light could be bent by a gravitational field. In this letter, Einstein explains to Hale his prediction of how light from a distant star could be deflected as it passed near the massive Sun. Einstein got to know Hale well during his visits to Southern California in the early 1930s.
Hale Collection, Box 154, The Henry E. Huntington Library and Art Gallery, San Marino, California

Fig. 45
Henry Huntington in
front of bronze
library doors,
San Marino, 1915.
The massive doors
display images of
important printers
and publishers of
the fifteenth and
sixteenth centuries.
Collection
Huntington
Institutional Archives,
The Henry E.
Huntington Library
and Art Gallery,
San Marino,
California

estate to a political and public agency. Hale outlined the idea of making the estate into a center for scholarship in the humanities. "Some of your suggestions are most excellent and I will take them under consideration," Huntington replied.[2]

2. Hale to Huntington, 17 April 1914, and Huntington to Hale, 20 April 1914, Mt. Wilson Papers, Hale Collection, Huntington Library.

Others had made the same suggestion to Huntington. For instance, Harrison Gray Otis, the publisher of the *Los Angeles Times*, made a similar plea at about the same time—to turn the estate and its contents into a public institution for the greater glory of Los Angeles. Meanwhile, Hale could not contain himself and kept at Huntington. He wrote him in May 1914: "The powerful attractions of your pictures and library have fired my imagination and set in motion a new train of ideas. I cannot help feeling that, with such rich and uniquely valuable material, it would be a very easy matter to make your collection of real international importance, without greater expenditure than you may already contemplate."[3] He then went on to detail his own experiences in starting major projects: the large observatories he had founded, the publications he had begun, and the organizations he had started. He sent a copy of the letter to his friend James Breasted. Impressed, Breasted noted, "This dream of an ideal project on the Pacific may be but the initial chapter in the spiritual development of the Pacific coast and . . . this letter of yours may someday be regarded as the historic germ out of which it has all grown."[4]

3. Hale to Huntington, 11 May 1914, Hale Collection.

4. Breasted to Hale, 19 May 1914, Hale Collection.

Huntington, always busy with travel to New York during the summer (Arabella preferred to spend summers in New York or Europe), did not reply until October. It was succinct but promising. "Your letter of May 11 reached me as I was sailing for Europe and during the summer I have given the suggestion some thought. I am not ready to reply, but it is quite possible that you have planted a seed."[5]

5. Huntington to Hale, 5 October 1914, Hale Collection.

The two men began to see each other more frequently. Hale was also an increasingly avid book collector, and he obtained a number of important astronomy books, many of which now reside at the Huntington Library. In 1919 Hale asked to serve on the library's first five-person board of trustees. He continued to agitate for long-term financial support and an endowment for the institution. As one of its trustees, he felt he had a significant voice in the library's future. "I do not think Mr. Huntington realizes all that I have in mind," he confided to Breasted.[6] Ultimately he succeeded in persuading Huntington to leave his millions for permanent support of the library, gardens, and art collections.

6. Hale to Breasted, 29 August 1925, Hale Collection.

Hale was certainly not the sole influence on Henry Huntington's decision to found a permanent public and scholarly institution and support it financially. He was one of the most critical and persuasive influences on Huntington, however, and the legacy left by the tycoon after his passing.

Hale's other great contribution to the cultural landscape of Southern California was the development of the California Institute of Technology. Founded in 1891, its precursor, the Throop Polytechnic Institute, was a small, poorly funded liberal arts undergraduate college with uneven standards, graduating two or three students annually with bachelor's degrees. Shortly after arriving in Pasadena, Hale took an interest in the small school. His neighbor Charles Frederick Holder was one of the college's trustees, and the two men often discussed the school. One night in early 1907, Hale suggested to Holder that the Throop Institute be made into a top-notch scientific institute. He urged the development of a research program, "not merely training," as he noted.[7]

7. Hale to Pritchett, 8 May 1907, Hale Collection.

Such a change would mean a reshaping of the school's entire character, dropping the elementary grades that were being taught, as well as nursing courses, and concentrating on fundamental science and several branches of engineering, as well as "adequate instruction . . . to all students in the humanities."[8] Hale was not talking idly. He was an increasingly respected scientist, and the previous year he had turned down the presidency of Massachusetts Institute of Technology, deciding instead to stay at Mount Wilson. Holder caught Hale's enthusiasm quickly and discussed the suggestions with other Throop trustees. "All of the Trustees," wrote board member S. Hazard Halstead, "are greatly pleased at the interest in this subject which you have expressed to several of the members and at your willingness to aid us in this matter."[9] The trustees quickly decided to offer a degree in electrical engineering, which they hoped to make the finest in the country, as well as agreeing generally that the Throop Institute should become a technical school of great quality.

8. Ibid.

9. Halsted to Hale, 29 April 1907, Hale Collection.

Hale was soon named to the Throop board of trustees, and the school was steadily redirected onto its new course. By mid-1909 $70,000 had been raised, plans for a new campus were being drawn up by architects Elmer Gray and Myron Hunt, and the number of students reduced from the original five hundred to a small but much more elite thirty-one. "This contraction of numbers in the face of a

Fig. 46
Staff and researchers
on Mount Wilson
(left to right):
H. L. Miller,
C. G. Abbot,
George Ellery Hale,
L. R. Ingersoll,
Ferdinand Ellerman,
W. S. Adams,
E. E. Barnard, and
C. S. Backus
Collection The Henry
E. Huntington
Library and Art
Gallery, San Marino,
California

great expansion of plan was probably the boldest step ever taken by an American educational institution," one Southern California publication remarked several decades later.[10] It was an audacious plan, but it worked, and the school's reputation grew.

       Hale spent a great deal of time and energy on his pursuits, and he had many interests, which he approached with a ferocious passion and enthusiasm. For his work in founding and overseeing the critical early years of two of the country's greatest observatories, the Mount Wilson Observatory and the Palomar Observatory, he is recognized as one of the pioneers of twentieth-century astronomy and astrophysics. He also played a vital but lesser-known role, however, in the founding of two of Southern California's most important cultural and educational institutions. That legacy continues today.

Dan Lewis is curator of American historical manuscripts at the Huntington Library, San Marino, California.

10. *Six Collegiate Decades: The Growth of Higher Education in Southern California* (Los Angeles: Security First National Bank of Los Angeles, 1929).

Kandinsky, Schoenberg, Einstein Constructive Modernism *before* World War II

If one were asked to name the most influential figures of the twentieth century, one would have to include the Russian-born Wassily Kandinsky (1866–1944), the Austrian-born Arnold Schoenberg (1874–1951), and the German-born Albert Einstein (1879–1955). The influence of all three extended far beyond their own particular fields: fine art, music, and science, respectively. They also have in common their escape from Nazi Germany and their death in exile: Kandinsky in France, Einstein and Schoenberg in the United States (Princeton and Los Angeles). Furthermore, they share a similar fate: although they changed the direction of art, music, and science, their work today, in the so-called postmodern era, is undergoing new critical examination. Kandinsky and Schoenberg no longer serve as models for many painters or composers, yet no one questions the importance of their contribution to the modern tradition.

Einstein's revolutionary theory of relativity made many aspects of traditional physics obsolete and opened the way for new discoveries, leading ultimately to the splitting of the atom. For Kandinsky and Schoenberg, the goal was to create a new, "constructive" art[1]—for Kandinsky in the spirit of the architecture and painting of the Italian Renaissance and for Schoenberg in the tradition of Johann Sebastian Bach. Since these traditional forms of art and music had already reached their zenith (for Schoenberg, in the music of Richard Wagner and Gustav Mahler, for instance), however, new forms and methods had to be found. Both Kandinsky and Schoenberg went through a metamorphosis before reaching their "final" destinations. Einstein, by contrast, published his first revolutionary theory when he was only twenty-six years old.

Fig. 47 (opposite)
Wassily Kandinsky
Pressure from Above, 1928
Watercolor and ink on paper, mounted on paper; 18 1/8 x 24 5/8 in. Norton Simon Museum, Pasadena, California, The Blue Four Galka Scheyer Collection, 1953

1. "Constructive" in this context refers to positive creativity and structure within modernism, as opposed to the distorted traditional forms and structures that Kandinsky saw in Picasso's cubism and Salvador Dalí's surrealism.

Fig. 48
Wassily Kandinsky,
1917
Photograph with
dedication on the
back to Schoenberg
Collection Arnold
Schoenberg Center,
Vienna, Austria

Before Kandinsky began to separate art from nature, leading to "abstract" and later "concrete" art, his paintings were based on Russian folk art, French impressionism, and Jugendstil (art nouveau). Nevertheless, in his impressionistic works he employed color for its strong visual effects and rhythms, so that it soon became more important than landscape, which was still quite often his subject. It was a period of "free reality," which ultimately led, in 1910, to the first painting in which Kandinsky freed art from nature completely. During this period he also wrote one of his most important theoretical works, "Concerning the Spiritual in Art"; founded the New Artist's Association in Munich; and was the cofounder of the expressionist artists' group The Blue Rider, which published an almanac of the same title. In this period Kandinsky was going further than most of his contemporaries in using color and form in an abstract, nonobjective manner.

The next important phase in Kandinsky's career was from 1920 to 1924. During World War I he moved via Switzerland back to Russia, where, after the Revolution, he became an influential figure. The period in which expressionistic and abstract artists were able to work freely under the Bolshevik regime was unfortunately only very brief, however, and in 1921 Kandinsky went back to Germany and became a professor at the Bauhaus in Weimar. At this time his art featured bundles of lines, points, circles, and triangles. In his next period, between 1925 and 1928 (the Bauhaus had moved to the German city of Dessau), he focused more and more on round, concentric forms.

In 1926 Kandinsky wrote another very important essay, "Point and Line to Plane," and his abstract period then led directly to the period of "concrete art" (which later also included "concrete poetry," developed by Eugen Gomringer [b. 1925]). For Kandinsky, the term concrete painting referred to works whose "material" components (color, circles, lines, points) were concrete in themselves and completely independent from the world of nature. And these components were used in a very conscious way; they were not merely the expression of a coincidental dreamworld, which was the case, for instance, in his first abstract painting, with its heavy color spots and shapes. Furthermore, this concrete or romantic painting ("romantic" in the sense that it echoed the demand of philosophers and authors of the German Romantic period for totally free fantasy in using words and other creative elements in art) not only was born out of the imagination but also—and this is very important—was considered to be highly constructive. Despite the emphasis on free fantasy, Kandinsky's "concrete paintings" (and later Gomringer's "concrete poetry") were not chaotic or anarchistic, but were based on positive systems, which the artists believed would open up a new order for the future. As a result, Kandinsky's "abstract" paintings became more and more geometric. Indeed, even in his earlier, more dynamic and dramatic "abstract" and expressionistic paintings, we see a tendency toward an inner (even geometric) order.

Fig. 49
Wassily Kandinsky
Small Worlds II,
1922
Lithograph in four
colors;
10 1/16 x 8 3/4 in.
Norton Simon
Museum, Pasadena,
California, The Blue
Four Galka Scheyer
Collection, 1953

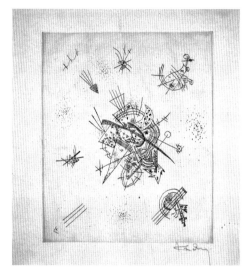

Kandinsky's use of colors and forms was also influenced by Johann Wolfgang von Goethe's *Theory of Color* (1810), a profound work in which the great poet demonstrated, among other things, the subjective perception of colors by the human eye. And yet, Goethe argued, despite this subjectivity and psychological influence (red creates the feeling of power, green the feeling of hope, etc.), all colors represent order. It is harmony and order we find, even in the appearance of colors when an object is viewed through the displacement of a prism (a central observation in Goethe's investigation of colors), and the consistent order of the colors creates "constructive" forms. (Some of Goethe's examples look like "abstract" paintings by Kandinsky or by concrete artists of more recent times, such as Richard Paul Lohse.)

It came as no surprise that Kandinsky, who became more and more obsessed with the "constructive" force of his concrete art, rejected the "new objectivity" in art and literature after World War I as "superficial" realism and also attacked the cubists, especially Pablo Picasso. For Kandinsky, Picasso's cubism did nothing more than distort traditional images of nature and the material world and was therefore destructive. He included in these attacks even Picasso's 1937 masterpiece *Guernica*, writing to his agent, Galka Scheyer, in Los Angeles: "Future generations will see with great interest, how our destructive world finds its reflection in the art of today," and refers to *Guernica*, with its loose (and almost grotesque) but still recognizable references to war.[2]

Similar to Kandinsky's evolution from his early paintings influenced by Russian folk art to a new "constructive" art was the metamorphosis of Arnold Schoenberg. We refer here to his music and will not discuss his paintings, which almost all belong to an early, expressionistic period. We also will not discuss the dispute between Kandinsky and Schoenberg concerning remarks by Kandinsky that Schoenberg viewed as clearly anti-Semitic. After this short-lived dispute, the two came to admire each other's art and theories.

2. Kandinsky to Scheyer, 2 October 1939, Norton Simon Museum, Pasadena, California. Excerpts from Kandinsky's correspondence with Scheyer were first published in Cornelius Schnauber, "Verehrte Frau Minister," radio play (9 August 1980), and in the essay of the same title, published in *Neue Zürcher Zeitung*, 10 August 1980.

Fig. 50
Wassily and Nina Kandinsky, Jertrud and Arnold Schoenberg, Portsdach, 1927 Collection Arnold Schoenberg Center, Vienna, Austria

Fig. 51
Twelve-tone row
chart of
Schoenberg's
Fourth String
Quartet, Opus 37
Collection Arnold
Schoenberg Center,
Vienna, Austria

3. This definition was
used by Schoenberg
in several essays and
letters. See Cornelius
Schnauber,
"Perception and
Apperception of
Dissonances in
Twelve-Tone Music,"
in *Journal of the
Arnold Schoenberg
Institute* 12 (June
1989): 5–21.

Schoenberg began his major compositions under the influence of Johannes Brahms's "neoclassic" structures and in the tradition of Richard Wagner's late Romantic style, with tonal and dissonant chords, overloaded polyphony, and stretched tonality (for instance, Schoenberg's monumental, expressionistic *Gurrelieder*, 1900–1911), which then led to "free atonality" (exemplified by *Pierrot Lunaire*, 1912). Even though we can already find in these compositions of free atonality a recognizable "geometric" order, as in Kandinsky's early expressionistic and abstract paintings, it was for Schoenberg not yet enough order. His goal was to place free atonality, with its dominating dissonance, into a new, "constructive" order. So between 1918 and 1920 he created the "method of composing with twelve tones which are related only with one another,"[3] which is a highly structured and sophisticated method. Explained in a simplified way, a "twelve-tone" composition begins with the establishment of a definite series in which each of the individual twelve half-tones (or "pitch classes") within an octave are handled equally (there are no dominating chord

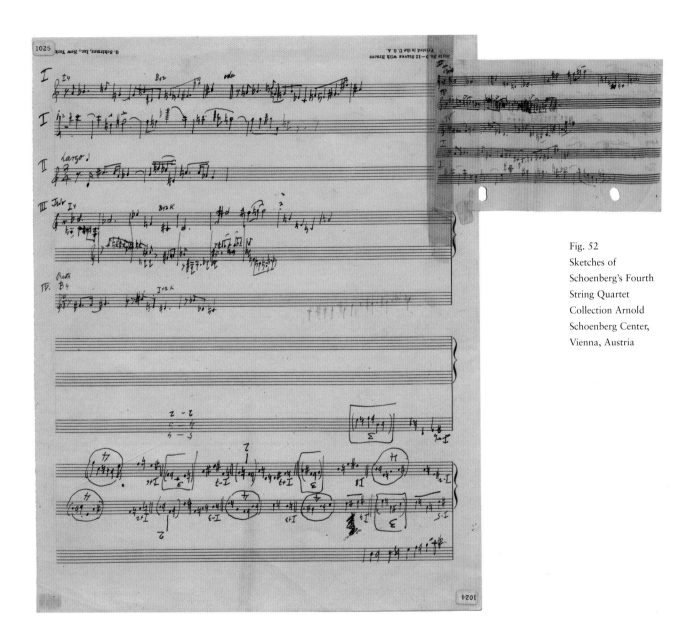

Fig. 52
Sketches of
Schoenberg's Fourth
String Quartet
Collection Arnold
Schoenberg Center,
Vienna, Austria

types, such as the triad) and are expanded through a specific system. In principle, no tone can be repeated until the others have been sounded. This applies to the horizontal progression as well as to the vertical harmonic structure (chords). Therefore, the twelve tones are in a particular order (a "row"), which has four basic forms: prime, inversion, retrograde, and inverse retrograde. And since each one of these forms can be transposed eleven times, there are forty-eight forms of the original row.

Fig. 53
Arnold Schoenberg at
the blackboard,
1940s
Collection Arnold
Schoenberg Center,
Vienna, Austria

No matter how well we understand this complex method of composing in the context of this brief essay, Schoenberg considered his new method—as Kandinsky considered his "concrete painting"—to be highly constructive. And as Kandinsky saw in Picasso the destructive artist, Schoenberg saw in the compositions of his German contemporary Richard Strauss a revolutionary, "anarchistic" music, since Strauss broke down tonality "destructively," disrupting it only episodically and in the end merely masking it with dissonance. Furthermore, while Strauss, in his most dissonant works, *Elektra* and *Salome*, powerfully developed the dissonances but then tonally returned them from the overcharged frenzy of sound back to the familiar, Schoenberg deliberately attempted to impart self-sufficiency to his dissonances as part of his system of constructive rules and structures.

Fig. 54
Schoenberg outside his house, Brentwood, Los Angeles, 1940s
Collection Arnold Schoenberg Center, Vienna, Austria

There is another dispute that should perhaps be mentioned here. In *Doktor Faustus* (1947), one of the German author Thomas Mann's most important novels, the main character is shown developing Schoenberg's cerebral twelve-tone music during a fit of nerves and migraines brought on by syphilis. This chapter of the novel led to angry correspondence between the two great exiled artists, who were both living in Los Angeles at that time. Most of the dispute concerned superficial matters, but the essence was that Mann portrayed the creation of twelve-tone music as an outgrowth of illness and a destructive and decadent period of European history (the years before World War I), while Schoenberg saw in it the very breakthrough to an ordered future.

Fig. 55
Albert Einstein lecturing at the
California Institute of Technology,
early 1930s
Collection The Henry E. Huntington
Library and Art
Gallery, San Marino, California

Albert Einstein's contribution to physics is known mostly through his theories of special and general relativity. The most momentous consequence of Einstein's theories of relativity is that they revised our traditional concept of space and time. Time is no longer defined in relation to the turning of the earth but in relation to the speed of light. Space and light are linked and create four-dimensional space, and time itself (in the theory of general relativity) becomes the fourth dimension. This modified concept of space and time "entailed that the length of a metre rod and the duration of a second on a clock depend on the state of motion of the observer with respect to the instruments. For an observer a moving metre rod will look shorter and the seconds on a moving clock will look longer."[4] Time loses its absoluteness (only the speed of light in a vacuum remains constant), and the "state of motion of the observer" becomes crucial. This order within new subjectivity, or "relativity," can be seen to be related indirectly to Kandinsky's and Goethe's observations of colors and to Schoenberg's new "relativity" of the twelve tones in our traditional octave system. And despite the word *relativity* in the titles of Einstein's two best-known works, his discoveries are highly constructive. Without them, we would still be in the technical and scientific world of the nineteenth century.

4. *Encyclopaedia Britannica*, 1964 ed., s.v. "Einstein, Albert."

In conclusion, one can say that only for the superficial observer is there the illusion of decadence in Kandinsky's concrete art (which the Nazis called degenerate), chaotic dissonance in Schoenberg's twelve-tone music, or insecurity within our earthly existence in Einstein's theories of relativity. Upon close examination, and according to these three great men themselves, the systems they developed represent a new order in art, music, and physics, respectively. Furthermore, this constructive modernism was in astonishing contrast to historical developments during the first third of the twentieth century, which Thomas Mann described in his novel *Doktor Faustus* as the beginning of the political and cultural downfall of Germany and Europe, culminating in the Nazi dictatorship and World War II.

Cornelius Schnauber is director of the Kade Institute for Austrian, German, and Swiss Studies at the University of Southern California, Los Angeles.

Contemporary Art
and the Cosmos

Since the early twentieth century artists have been absorbed with themes or forms associated with the cosmos. Abstract artists in particular have been attracted by the formal possibilities of the circular or rectilinear geometry and universal mathematical order that, since Pythagoras in the sixth century B.C., have been associated with the cosmos. For many, this subject has spiritual implications as well, as a metaphor for the infinite or unknowable. New astronomical discoveries have been an enduring source of fascination, and after World War II, when space exploration became a reality, artists took inspiration from space expeditions and the photographic images resulting from them. Others were intrigued by the potential of technology to provide means to explore the cosmos and connect us to its order and vastness.

Fig. 56
Rockne Krebs
(b. 1938)
The Green Hypotenuse, 1983
Eight-mile-long laser installation from Mount Wilson to California Institute of Technology
Courtesy of the artist

In the 1910s the Italian futurist Giacomo Balla expressed that movement's fascination with the machine and universal dynamism in such paintings as *Mercury Passing before the Sun Seen through a Telescope* (1914). The formal geometric innovations of cubism and fauvism in France were at times also connected with the celestial. Robert Delaunay's Circular Forms series of 1912 suggested both the infinite depth of space and the spherical forms of the planets.[1] The Russian constructivists often emphasized the physicality of materials and new artistic forms. In what may be that movement's single greatest work of art, Vladimir Tatlin's

1. Constance Naubert-Risier, "Cosmic Imaginings: From Symbolism to Abstract Art," in *Cosmos: From Romanticism to the Avant-garde*, ed. Jean Clair (Montreal: Montreal Museum of Fine Arts; Munich: Prestel, 1999), 220.

proposed *Monument for the Third International* (1919), the new communist political order in Russia was linked to the celestial order. Each of the monument's three suspended elements was intended to revolve once a day, once a month, or once a year. The surrealist movement, which appeared after World War I, looked inward to the unconscious as a source for art. In Joan Miró's and Alexander Calder's organic forms, one finds the geometry of circles and lines employed to express truths about the creation of the world or the order of the cosmos.

After World War II the simple geometry of Mark Rothko's and Barnett Newman's large-scale abstract paintings seemed to refer metaphorically to the order of the cosmos. The 1950s saw the rise of a mass culture dominated by images conveyed by glossy magazines and television. Around 1954 Robert Rauschenberg began to unify these images and objects with the materials and forms of art in his Combines. The visual language of 1960s pop art, coming out of Rauschenberg's work and the emblematic images of Jasper Johns, embraced the most common forms and objects of the age of consumerism.

2. Douglas M. Davis, "Rauschenberg's Recent Graphics," *Art in America* 57 (July–August 1969): 90.

Throughout his career Rauschenberg has embraced an exceptionally wide variety of mass images, including those of space exploration. Having transformed painting, drawing, and sculpture in the 1950s, in 1962 he began to explore lithography, working at distinguished print shops such as U.L.A.E. on Long Island and Gemini G.E.L. in Los Angeles.[2] In the late 1960s, after watching a manned launch at Cape Canaveral, he decided to do a series of lithographs celebrating the astonishing achievements of NASA's Apollo mission to the Moon (see fig. 57). The title of the Stoned Moon series is a double pun, referring both to the act of putting images of this historic event on lithographic stones and to the metaphoric "high" of putting a man on the Moon. Rauschenberg's extraordinary abilities to combine images allowed him to convey the poetry, and not just the technological wonder, of the lunar mission, for example, depicting the swamp from which the immense rockets were fired and events happening thousands of miles away in the Apollo space capsule in the same lithograph. In this series, one of the most important events of the twentieth century was celebrated by one of the period's greatest imagists.

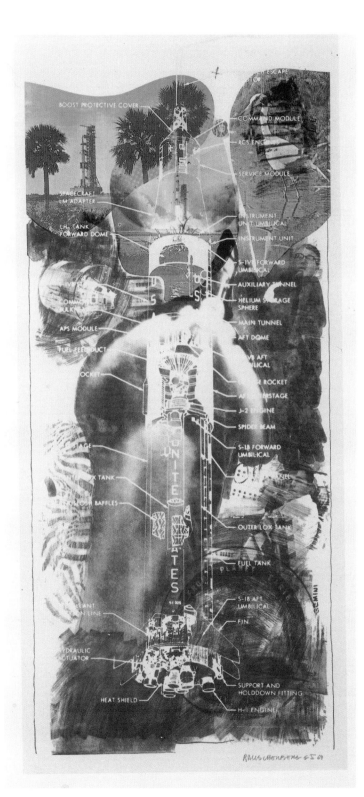

Fig. 57
Robert Rauschenberg
(b. 1925)
Sky Garden, 1969
Color lithograph and
silkscreen; 89 x 42 in.
Courtesy Gemini
G.E.L., Los Angeles

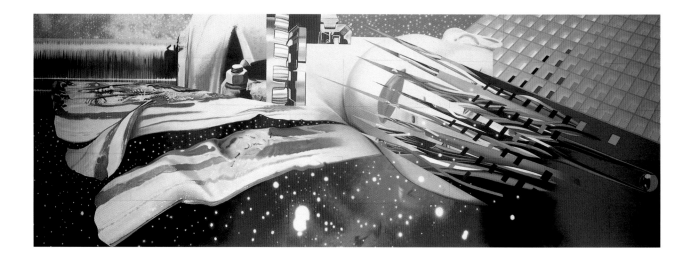

Fig. 58
James Rosenquist
(b. 1933)
Star Thief, 1980
Oil on canvas;
204 x 552 in.
Collection Art
Enterprises, Ltd.,
Chicago, Illinois

3. Constance Glenn,
   "Time Dust: The
   Essay," in *Time Dust:*
   James Rosenquist:
   Complete Graphics,
   1962–1992 (New
   York: Rizzoli, 1993),
   108.

4. Ibid.

5. Ibid., 124.

James Rosenquist, who became associated with the pop movement in the early 1960s, has also explored celestial imagery. He developed an approach to monumental painting inspired by the immense scale of the billboards that he had painted early in his career. His paintings drew upon the graphic simplicity of advertising, which gave them a formal clarity, yet his method of combining images endowed them with a poetic dimension. For Rosenquist, the late 1970s and early 1980s, as Connie Glenn has noted, "proved to be a period of gestation that would spawn a completely unexpected methodology for exploring space and depth perception" in his work.[3] This new direction was initiated by the conception and completion of the forty-six-foot painting *Star Thief* (1980; fig. 58). According to Judith Goldman, "The depicted space appears infinite, as unseeable and unknowable as we once believed outer space to be."[4] A number of the later Rosenquist graphics further explore the formal innovations and themes of *Star Thief*. In the 1991 print series Welcome to the Water Planet, on which Rosenquist collaborated with master printer Kenneth Tyler, the artist combines visual theatrics with virtuoso papermaking and printmaking, "once again set[ting] out to explore the mysteries of time and space from the vantage point of his ongoing concern over the future of our planet."[5]

Since the late 1950s Ed Ruscha has also explored the possibilities of pop imagery, often taking emblematic words or logos as his subjects. Language has always been important to Ruscha, who was trained in graphic design. He has made us understand that words are signifiers and forms at the same time, and his works have often explored the almost surrealistic outcome of presenting both their identities concurrently.

In the late 1970s Ruscha made a number of monumental horizontal paintings that ostensibly depicted the edge of our planet with small words disposed at curious points in the "sky." Rather than exploring so-called regional themes, as in his

Fig. 59
Edward Ruscha
(b. 1937)
Various Places, 1979
Oil on canvas;
22 x 80 in.
Collection Burroughs
Corporation

well-known images of Los Angeles, he began to deal, almost literally, with universal ones. And in paintings like *It's a Small World* (1980), the reality of the human position in the larger scope of things is more explicit. For this exhibition he has created a new series of artworks investigating our relationship to the universe. Lost in space, lost in language, Ruscha's works, as Carrie Rickey has noted, "encompass that . . . limbo which is not the certainty of disbelief but rather the anxiety of doubt."[6]

6. Carrie Rickey, "Ed Ruscha, Geographer," *Art in America* 70 (October 1982): 91–92.

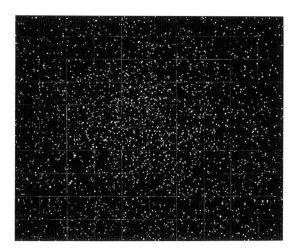

In the 1960s and 1970s Vija Celmins revitalized the genres of still life, landscape, and marine and sky studies in small and surprisingly traditional paintings and drawings.[7] She became fascinated with images that implied vast reaches of space. In black-and-white drawings requiring extraordinary effort and technical mastery, she depicted slightly differentiated areas of low waves, flat desert landscape, and the starry sky. As an artist who developed in the late 1960s, Celmins was very aware of her works of art as flat objects, and she emphasized their uniformity of imagery, formal texture, and surface. In their emptiness and lack of specificity, her images seem outside time and space.

From 1982 through 1985, working at Gemini G.E.L., Celmins created a number of prints depicting stars, which combined aquatint, mezzotint, drypoint, and etching (see figs. 60, 61). In some of these prints she associated images of the night sky with other subjects. Most telling are several prints that pair an image of the starry firmament with the large freestanding motorized rotating piece *Rotary Glass Plates (Precision Optics)* (1920) by Marcel Duchamp. Though her selection of the image of Duchamp's work speaks of his—and her—use of found objects (or images) as the source of art, for Celmins, Duchamp's circular machine clearly evokes some extraordinary celestial mechanism.

Fig. 60 (top)
Vija Celmins (b. 1938)
Strata, 1982
One-color mezzotint; 29 ½ x 35 ¼ in.
Courtesy Gemini G.E.L., Los Angeles

Fig. 61 (bottom)
Vija Celmins (b. 1938)
Concentric Bearings A, 1985
Two-color aquatint/ photogravure;
23 ⅞ x 18 ½ in.
Courtesy Gemini G.E.L., Los Angeles

7. Lane Relyea, "Earth to Vija Celmins," *Artforum* 31 (October 1993): 58.

8. Lucinda Barnes, "Observatory/ Territory," in *Kim Abeles: Encyclopedia Persona, A–Z*, ed. Karen Moss and Scott Boberg (Los Angeles: Fellows of Contemporary Art, 1993), 70.

Linda Connor's photographs reveal our connections with nature and, through the art or architecture of ancient cultures, to the sacred. Time is another theme in her works—the time of history, the eternal character of the gods, the cyclical time of nature, and the cycle of changing light so crucial to photography. The equipment and film that Connor uses contribute significantly to the sense of quiet and timelessness in her work. First, she employs an old view camera, which makes photography a slow, arduous process; second, her locations often require long exposures; finally, she makes her prints the size of her negatives on printing-out paper, which may require several hours of sun exposure before the image is ready to be fixed.

In 1985 Connor discovered the glass negatives of photographs taken through the telescopes at Lick Observatory in Northern California more than a century earlier. The sepia tones of the printing-out paper evoke the age of the glass plates as well as the celestial objects in their images. When Connor exhibits her photographs, she intuitively orders them to create unexpected and often mysterious connections. In this exhibition, in which Connor's Lick photographs are installed with her images of religious architecture from other cultures, the result is an implied encyclopedia of the celestial, in which the spirit and the heavens are associated and made one (see figs. 62, 63).

Fig. 62
Linda Connor
(b. 1944)
Fountain Head, Angkor, Cambodia, 2000
Photographic negative on printing-out paper; 10 x 12 in.
Courtesy of the artist

Fig. 63
Linda Connor
(b. 1944)
July 1895, 1997
Glass photographic plate on printing-out paper; 10 x 12 in.
Courtesy of the artist and University of California Observatories, Lick Observatory

Fig. 64
Kim Abeles (b. 1952)
Illuminated
Manuscript, 1984
Ink, acrylic, and
pencil on paper;
51 x 78 ½ in.
Collection of Judy
and Stuart Spence

In her works about the cosmos, Kim Abeles reveals a comprehensiveness of approach that is characteristic of all her installations and artists' books. Since 1979 she has worked in series, gathering extensive information on various subjects. She assembles her art from found materials or handmade objects, but these are always in the service of her social, political, and environmental concerns. Abeles's installation *Observatory/Territory* (see fig. 64) was created and "performed" during her 1984 residency at Hand Hollow Foundation, George Rickey Workshop, in upstate New York.[8] She built and worked in an outdoor sculptural environment for a period of time between new moon and full moon to chart the paths of the Sun as well as the Moon. Abeles placed five sheer fabric panels on the sculpture each day. After the positions of the sun and moon were recorded, the fabric was taken to the studio, where the information was duplicated in acrylics for the Sun images and cutout shapes

for the Moon. In this piece, Abeles connects herself, and the viewer, to the constantly changing spatial relationship of the Earth to other celestial bodies and, by implication, to the cosmos itself.

    From early in her career, Dorothea Rockburne has pursued artistic ideas that, for her, are most appropriately explored in the context of mathematics and science. In the 1970s and 1980s she focused on a wide range of mathematical theories. At the same time she was drawing inspiration from Italian Renaissance masters such as Giotto, Piero della Francesca, and Ambrogio Lorenzetti, as well as "the spatial and coloristic dissonances of later Mannerism."[9] As the 1980s continued, she immersed herself in astronomy and astrophysics, referring in her art to innovators in these fields, from Copernicus to Stephen Hawking, as well as to images

Fig. 65
Dorothea Rockburne
Wanderers, 2000
Aquacryl, silver
Caran D'Ache,
One Shot, Deka
white enamel on
handmade paper;
12 1/8 x 17 1/4 in.
Courtesy the artist
and Lawrence Rubin
Greenberg Van Doren
Fine Art, New York

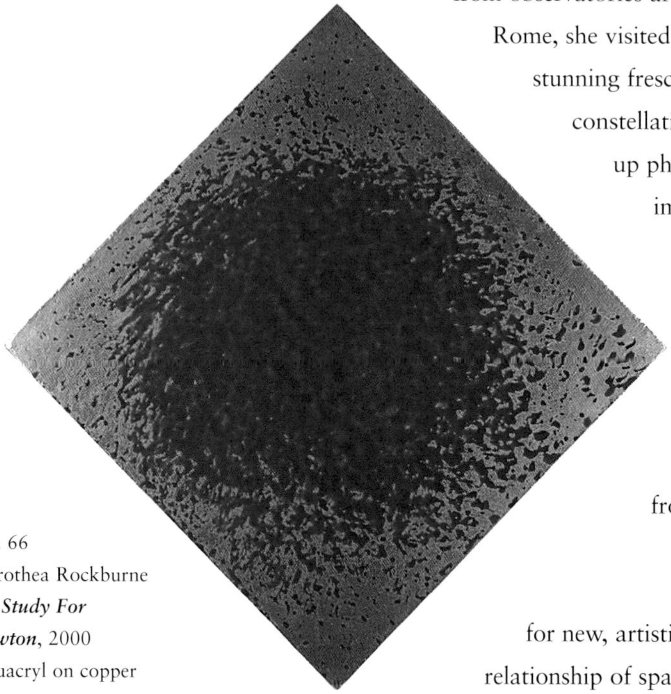

Fig. 66
Dorothea Rockburne
*1st Study For
Newton*, 2000
Aquacryl on copper
foil; 14 x 14 in.
Courtesy of the artist
and Lawrence Rubin
Greenberg
Van Doren Fine Art,
New York

from observatories around the world. In 1991, while in Rome, she visited a seventeenth-century villa that had a stunning fresco showing the planets and their orbits, constellations, and stars. A collage including cut-up photos of the fresco has served as the impetus for ten years of her astronomical drawings, in which the investigation of abstraction in her earlier work is unified with the geometry of the cosmos. Much of this work is inspired by the extraor-dinary color images transmitted from distant space.

Rockburne has constantly searched for new, artistically satisfying ways to depict the relationship of space to time. As one critic has noted, her art "affirms the age-old concept that a singular, simple beauty governs the structure of the universe, that physical laws are also spiritual, and that a profound elegance shapes both."[10]

Rockne Krebs connects with the vastness of space through technology. When he began experimenting with the laser in 1967, he realized that it "had a visual vocabulary of its own."[11] It defined space by lines of a material that was completely insubstantial, light. Line automatically implies direct distance, and by the early 1970s he was creating miles-long installations of lasers, often reflecting them off mirrors to increase their dimensions. These weightless sculptures evoked the individuality and metaphoric meanings of their outdoor locations. In 1983, for Baxter Art Gallery at the California Institute of Technology, using light from a single argon laser, Krebs connected Mount Wilson to the wall of Caltech's Beckman Auditorium more than seven miles away (fig. 56). He allowed the laser beam to spread, so that by the time it passed through the changing air currents to the white exterior of the auditorium, it was a constantly altering "painting" of green light. Although he did not know it, there was an appropriateness to linking an area near the Mount Wilson Observatory (where Edwin Hubble developed the modern theory of the universe) to

9. Lilly Wei,
   "Watching the Skies,"
   Art in America 88
   (April 2000): 128.

10. Ibid., 132.

11. Rockne Krebs,
    interview with the
    author, 3 November
    1982.

the scientific institution associated with some of the most important and unique space exploration of the century. The exterior laser installation that Krebs has developed for *The Universe* reflects a similar awareness of site.

The idea of using beams of sunlight, which became Krebs's second important medium, occurred to him as early as 1969. Since the source of sunlight was millions of miles away, it was even less of a material presence than the laser and had much broader conceptual implications in terms of the space encompassed by his art. The exhibition includes several drawings of his proposed or completed sunlight works, which speak about our, and the artist's, relation to time and space in the broadest context.

To Carl Cheng, art "results from an attitude that . . . humans are not adversaries of nature but part of nature. . . . Art and science are two aspects of contemporary human nature, which incorporate our ideas, beliefs, discoveries and inventions."[12] In Western culture nature is often seen as the opposite of culture and technology. Cheng's aim is to link technology with Eastern spirituality in individual works of art and site-specific installations that encourage both intellectual thought and contemplation. He wants to work directly with the community, rather than being a gallery artist who makes art as a commodity.[13] He therefore focuses on public art, making works that are informed by his belief in the connection of nature and technology.

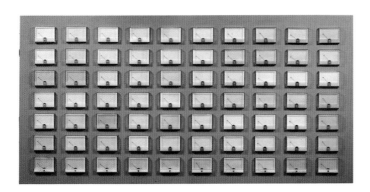

Fig. 67 (left)
Carl Cheng (b. 1942)
Conversation, 2000
Electronic hardware,
80 meters;
96 x 144 x 4 in.
Courtesy of the artist

12. Carl Cheng, artist's statement, 1999.

13. J. S. M. Willette, "Rite of Privacy: The Public Art of Carl Cheng," *Visions* 7 (fall 1993): 45.

Cheng's project for the Metro Rail Station in Redondo Beach, California, for example, acknowledges the proximity of high-tech firms as well as the Pacific Ocean (fig. 68). The lower part of the station evokes the undersea and underground, while the second-floor platform depicts astronomical information such as orbits of planets and a live NASA television transmission from spacecraft and

Fig. 68
Carl Cheng (b. 1942)
Metropolitan
Transportation
Authority Station,
Los Angeles: View of
revolving satellite
sculptures mounted
on top of elevator
shafts, 1998

satellites. Aware of the amount of wave energy that constantly bombards the Earth, Cheng has created an installation at the Armory Center for the Arts consisting of a wall of dials that he says is "inspired by the scientific exploration of the universe through radio telescopes and satellites that rely on cosmic energy not perceivable by humans except through instrumentation."[14] Formally, the result is a geometric grid of equipment that reveals information from radio waves coming from far away (fig. 67).

14. Carl Cheng, artist's statement, 2000.

In the history of art, there are particular periods that precede or find form for innovative developments in science. During the first half of the twentieth century there was an optimism about the modern age, as humankind attempted to explain and understand the physical realities of the universe through the revolutionary theories of figures such as Einstein or with data from new telescopes such as the one at Mount Wilson. In the last five decades the investigation of space— by humans or machines we have created—has changed human history and also the variety of imagery available to artists. Artists such as Rauschenberg and Rosenquist have been inspired by these images of the planets, stars, and galaxies. In the work of Celmins and Rockburne, the forms of abstract art are no longer divorced from what can be seen in the cosmos. Other artists, such as Cheng and Krebs, have been inspired by the ability of technology to enhance our understanding of the universe. Finally, in the work of artists such as Linda Connor, visual references to the celestial sphere convey our physical and spiritual relationships to the universe.

Jay Belloli is director of gallery programs at the Armory Center for the Arts, Pasadena, California.

A Sense
of Adventure

Fig. 69
Scene from **Contact**:
The young
Eleanor Arroway
after the camera
pulls back from its
exploration of the
universe.
Directed by
Robert Zemeckis for
Warner Bros.
Courtesy Warner
Bros.

Since the early experiments of Georges Méliès's *A Trip to the Moon* (1902), Holger Madsen's *Airship* (1917), and Fritz Lang's *The Woman in the Moon* (1929), several generations have used film as a medium for imagining alternative futures. In *Things to Come* (1936), *The Day the Earth Stood Still* (1951), *Blade Runner* (1982), and *Contact* (1997), each an iconic statement of its time, the future is in the hands of science and scientists; each film presupposes that technology is the engine of history, driving political, economic, and social change. Together the films describe a century of disenchantment with traditional authority, beginning with political, military, and economic authorities and extending finally to science. We move from a residual faith in the nineteenth century's ideal of human progress to postnuclear dystopias. But the films also describe a countertrajectory that moves from the authoritarian reign of pseudoreligious experts to a more modest collaboration between science and democracy. Through all the films runs a desire for exploration that still signifies our utopian longings.

Things to Come is a translation of H. G. Wells's 1935 nonfictional tract *The Shape of Things to Come*. Wells hoped to use the new medium of film to promote his "open conspiracy" of a global scientific and technical elite working together for world salvation. While muting the author's political critiques, the film remains a hymn to industrial progress and the triumph of technocratic order. Since the 1880s life in the British empire had been experienced as a continuous emergency, sharply punctuated by European contests for dominance at home and abroad. For Wells's generation, the expectation of war was a fact of everyday life. Wells placed his hope in science's ability to harness the technology of war for peaceful projects.

Fig. 70
Scene from *Things to Come* (1936): The Boss of Everytown (Ralph Richardson) argues with John Cabal, the Air Dictator (Raymond Massey). Roxana (Margaretta Scott), the Boss's mistress, stands next to the Boss. Directed by William Cameron Menzies for London Films (Alexander Korda, producer) Courtesy of the Academy of Motion Picture Arts and Sciences, Beverly Hills, California

In *Things to Come* the superiority of the new order over the old is associated with its command of new knowledge, expressed through powerful technologies: the airplane, the gas of peace, the underground city, and the great space gun. Indeed, the film reaches its highest pitches when human beings disappear from view. The death of the Mussolini-like Boss and the triumph of the Airmen introduce a sequential montage that records Everytown's transformation into a utopia. As giant machines drill out the new world, an energetic procession of industrial images—dynamos whirring, liquid metal pouring, sides of mountains exploding, large sheets of material being shipped from place to place—reveals the future as a great feat of civil engineering. The central problem of the film, the debate between progress and rest, is resolved by the visual vocabulary of technological utopianism. The debaters, Theotocopulos and Cabal, are reduced to near invisibility as the camera attends to the scale and spectacle of the Space Gun. Their magnified voices boom over images of moving machinery. The monumentality of the gun; the slow, inevitable movement of its pistons; and the sensuous spectacle of its firing justify the first space voyage and dramatize the artist's failure to see the future.

Despite Wells's socialist doctrine of progress, *Things to Come* records a curious tension between the egalitarian hopes evinced by the "brotherhood" of the Airmen and the reality of the Air Dictatorship. Even as it imagines "the unity of a common order and a common knowledge," the film presents the future as a win-or-lose proposition, not the collaborative product of all our knowledge. This tension is meant to be resolved in the figure of the scientist, who is at once prophet and messiah, earthly builder and political leader. Trust him, the film advises, and all our other ideals—democracy, equality, liberty—will be realized.

Just two decades later the optimism of *Things to Come* seemed laughable. With global holocaust a real possibility, *The Day the Earth Stood Still* looks to "outside" help to restrain humanity's destructive urges. The film was released in the spring of the Cold War as the euphoria that followed the Allied victory faded into fear for the future. During its production the Korean conflict raised the specter of a third world war. The film ends not with the glorious vision of a conquest of space, but with a stark choice: submit to peaceful coexistence or be destroyed. Again, international science succeeds where politicians fail. When Klaatu, the alien emissary, finds that he cannot communicate his message of peace to the governments and militaries of the world, he turns to its scientists. Professor Barnhardt is presented to us as "the greatest man in the world," at least since Abraham Lincoln. Klaatu's own greatness is demonstrated by his command of the language of mathematics, which allows him to communicate with Barnhardt. Collectively the world's scientists form an ideal community composed of a broad spectrum of races and cultures, bound together by curiosity and knowledge rather than dogma or ambition.

Fig. 71
Scene from
The Day the Earth Stood Still:
Klaatu (Michael Rennie) with Professor Barnhardt (Sam Jaffe). Directed by Robert Wise for Twentieth Century Fox (Julian Blaustein, producer) Courtesy of the Academy of Motion Picture Arts and Sciences, Beverly Hills, California

The film recommends that the world put aside aggressive competition and come together as one polity. Only then will we be fit neighbors for any galactic club that might come calling. And it offers the utopian technoscience of Klaatu's civilization as a powerful inducement for good behavior. But again there is a tension between good government and just government, oligarchy and democracy. If corrupt interests illegitimately control earthly power, the solution is neither to return the mechanisms of governance to the governed nor to make its institutions accountable to us. The solution is simply to put the right people in charge. And again a messiah figure rises to the challenge. But now human scientists no longer herald salvation; that task falls to Klaatu and his companion. Edmund H. North, the screenwriter, linked the extraterrestrial emissary to Christ through his alias, "Mr. Carpenter"; his miracles; and his resurrection. The scientists Klaatu assembles function as his disciples and are thus qualified to be our legitimate, albeit unelected, leaders. Doubts about the legitimacy of this political solution are evident in his assurance that although humanity will have to accept a universal order with an established set of rules and the police to enforce them, obeying those rules would "not mean giving up any freedom."

Fig. 72
Scene from *The Day the Earth Stood Still* (1951): The alien Klaatu is shot by the U.S. military. Directed by Robert Wise for Twentieth Century Fox (Julian Blaustein, producer) Courtesy of the Academy of Motion Picture Arts and Sciences, Beverly Hills, California

In *The Day the Earth Stood Still*, scientists remain distinct from corruptible political and economic interests. That distance disappears by 1984. Suddenly the world is in full color, and the future is bleak. Although made in 1982, *Blade Runner* was first released in a year made significant by George Orwell's 1948 dystopia. As an architect of that grim future, Ridley Scott developed the visual vocabulary that replaced the clean, freshly scrubbed surfaces of *2001: A Space Odyssey* (1968) with the smoggy, greasy interiors of his 1979 masterpiece, *Alien*. In *Blade Runner* Scott plays to our unease with authority figures, who have too often proven irresponsible or corrupt. He thus forecloses the possibility that science will provide a deus ex machina that will save us from ourselves.

In *Blade Runner* the scientific acquisition of knowledge is overwhelmed by the noise of a hypercharged consumer economy. The conquest of space is part of that noise, through a real estate scam that produces extraterrestrial suburbs for the fortunate. The master of this future is Eldon Tyrell, the hero-scientist corrupted by age and power. As the Second Empire decor of his bedroom indicates, he is the Frankenstein of the new world order. Tyrell represents a science that literally cannot

see the implications of its work, a limitation made brutally obvious when his eyes are crushed by one of his creations. With his death the quest for knowledge and faith in its liberatory potential seem to reach a dead end. But despite its hardboiled pessimism and its neo-noir evocation of corruption, the film offers hints of an enduring hope that there are still discoveries to be made and experiences to cherish. Its last word goes to Roy Batty, the replicant who kills his less than godlike "father." Batty's final speech begins, "I've seen things you people wouldn't believe," and he goes on to list astronomical wonders encountered off planet. He has seen into the structure of the universe, and in finding that order, he has awakened to himself and to his condition. *Blade Runner* thus combines pessimism about a fast-approaching commercial dystopia with older, utopian longings for an innocent, unmediated relation between self and world, knower and known, still figured by the scientific quest. In the film's fallen world, however, no one is untainted by what Rachel calls "the business"; hence, there is no one whose vision we can trust.

Fig. 73
Scene from
Blade Runner (1982):
Genetic designer
Eldon Tyrell
(Joe Turkel) of the
Tyrell Corporation
Directed by Ridley
Scott for Warner
Bros. (Ladd/Blade
Runner Partnership,
producers)
Courtesy Warner
Bros.

If *Blade Runner* is a visual parable recording our fears, then the opening sequence of *Contact* is a reminder that optimism and idealism are still expressed through the trope of science. As the camera pulls back from earth and takes us on a grand tour of our solar system, galaxy, and cosmological neighborhood, we are reacquainted with an awe that has remained consistent since *Things to Come*. The camera then leaves the Milky Way and pulls back through a sea of galaxies contained in the eye of a child. The eye, which is a symbol of surveillance and oppression in *Blade Runner*, is in *Contact* the portal to wonders, and the innocence of scientific knowledge and community are rediscovered in a child's sense of adventure.

Fig. 74
Scene from
Blade Runner:
The replicant
Roy Batty
(Rutger Hauer)
Courtesy Warner
Bros.

It is hard not to see this visual statement as the encapsulation of the last century of intellectual aspiration. It celebrates the excitement of exploration and curiosity unfettered by the "realistic" claims of commerce, territorial expansion, or defense. By ending at a ham radio station, the sequence harks back to the popular science magazines of the early 1900s, such as *Electrical Experimenter*, which catered to "wireless" enthusiasts and nurtured modern science fiction. While advocating the search for extraterrestrial intelligence, *Contact*, based on Carl Sagan's novel, also highlights competing visions of the value and purpose of science. The fundamental conflict, dramatized in exchanges between Eleanor Arroway and David Drumlin, is between science as a political and economic resource and science as an ideal pursuit of knowledge. Arroway is firmly in the tradition of the selfless medical doctors and engineers of *Things to Come*. She is the idealistic young radio astronomer whose only concern is the expansion of human knowledge.

While *Contact* brings us back to the characters of *Things to Come*, it eschews the traditional utopian plot. The link between intellectual and worldly interests, once severed, cannot be repaired. The authorized scientist—the white, male Drumlin—is exposed as an opportunist who cannot be trusted. It is Arroway, the scientist marginalized by all authorities, including those of her own profession, who heralds the future. The corruption that was once external to science has now become a part of its working order. The discipline is no longer a meritocracy; therefore it cannot represent the values of orderly and just government. *Contact* all but disowns the spirit of public service that has been part of professional science since the middle of the nineteenth century. In its stead appears a private faith in the personally transformative potential of an incorruptible love of knowledge for its own sake.

Almost, but not quite. In the final scene of the film the current of technological utopianism that has run through the twentieth century is renewed. As Arroway entertains the questions of a group of schoolchildren, we see that hope for the future is still invested in the figure of the scientist. Now, however, the scientist is a woman, a representative of the disenfranchised. She is not a master of the world handing on the mantle of absolute power, but a teacher hoping to pass on the values of intellectual curiosity. She encourages her students to find their own answers to life's questions. The children, in turn, are not the disciplined soldiers of an authoritarian technocracy, but the informed citizens of an emerging, meritocratic, and possibly democratic future. And the good order of that future rests on the egalitarian inclusion of all the races and genders they represent.

The resonance between this stubborn optimism and the history of radio astronomy points to the significance of science throughout the twentieth century as a repository of hope. Whether the next generation of filmmakers can imagine a universe worthy of the ideals we justify through science remains to be seen.

De Witt Douglas Kilgore is assistant professor of English and American studies at Indiana University, Bloomington.

Russell Crotty
A Diagram *of* Forces

Consider the bright star Sirius, located in the constellation Canis Major. Its waves of light ripple for eight and a half years across desolate expanses of black space before finally washing ashore on the retina of any earth creature caught gazing in its direction. Millions of nocturnal animals register Sirius's timeless presence, perhaps only as a momentary glance, and then go about the more pressing business of searching out sustenance on their planet's surface. Seeking to appease a different appetite, however, certain humans fix their gaze more intently, satisfying their curiosity as Sirius and a vast host of other shining points follow their beaten paths across the dome of sky. One of those observers, on any given night, is the artist Russell Crotty.

In Malibu, California, the Pacific Ocean delivers its own waves, churned up by storms and hurricanes far out at sea. These water waves travel outward from their source, pushed by wind and tide over vast distances toward the shoal waters where land emerges. Each wave is a flowing manifestation of physics as its crest accelerates relative to the rest of the swell and ultimately becomes unstable, toppling over on itself and squandering its remaining power by grinding bits of shell and rock into sand. Before the end of this journey, at the point where the approaching seabed causes each wave face to steepen until it's nearly vertical, one might, on any given day, find Russell Crotty tucked beneath a wave, pushed sideways and forward at the same time by its relentless procession, viscerally connecting with the primary forces that echo at last against the shoreline.

Fig. 75
Russell Crotty
Milky Way Northern Hemisphere, 2000
Ink on paper, mounted on Lucite sphere; diameter: 36 in. Collection of Ron and Ann Pizzuti, Columbus, Ohio

Fig. 76
Russell Crotty
*M13, Globular
Cluster in Hercules*,
1999
Ink on paper,
mounted on
Lucite sphere;
diameter: 24 in.
Collection of
Howard Rachofsky,
Dallas, Texas

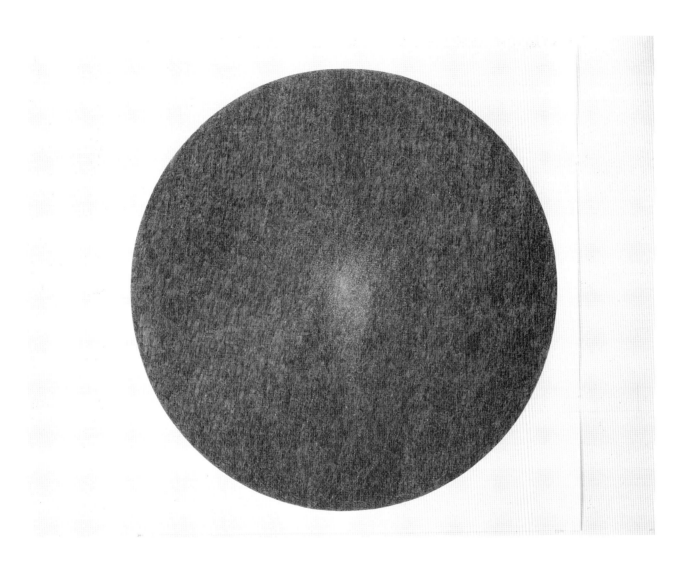

Fig. 77
Russell Crotty
*NGC 7789,
Galactic Cluster in
Cassiopeia*, 2000
Ink and pencil on
paper; 48 x 48 in.
Courtesy of Shoshana
Wayne Gallery,
Santa Monica,
California

Like John Muir, Crotty has strapped himself to a tree during windstorms, witnessing firsthand another of the forces that have sculpted the earth's natural geographic countenance—a process that brings into being those shapes and organisms biologist D'arcy Thompson referred to in 1917 when he described form as "a diagram of forces." But for Crotty the reward of riding out windstorms or surfing ocean waves is not in the extreme sportsmanship of putting himself at risk, but rather in the sublime act of connecting in a straightforward and unmediated way with nature.

So it is with Crotty and astronomy. On Solstice Peak in Malibu, 580 feet above sea level, up a dirt road from the hilltop cottage he shares with his wife, graphic designer Laura Gruenther, Crotty has built an observatory. A shedlike structure with a roll-off roof, it houses a ten-inch f/8 Newtonian reflector telescope on a massive equatorial mount. Armed with this and several other pieces of strictly analog stargazing equipment, he observes and diagrams by hand what he sees, training his eye to discern the subtle and revelatory differences that define incomprehensibly distant space objects. In this unlikely setting, he rides waves of light, peering into the past, navigating around the shuddering realization that these great masses of matter are actually *out there*, silently unfolding in time according to their own realities while we on earth go about our daily routines. The simple experience of looking and recording, so much a part of humankind's long and intimate relationship with the cosmos, is what drives Crotty's methodology and forges his art.

This is a poetic position, perhaps, and even a romantic one to some degree. But, decidedly unimpressed by popular

Fig. 78
The artist's Solstice
Peak Observatory

Fig. 79
The artist's studio,
summer 2000

culture's New Age antics, Crotty carefully avoids any spiritual implication that one might draw from his close connection with nature. What comes through in conversation is the profound awe with which he regards his subject matter and the minimalist's purity with which he pursues his research. Science—and the grandeur glimpsed between the lines of its investigation, a grandeur revealed precisely *because* of the way science undertakes its quest—is Crotty's working model. As a certified documentarian of the Association of Lunar and Planetary Observers, he is allowed to submit sketches of his observations to the organization's journal, sketches that must follow strict guidelines

for style and format. The rigorous protocol for these submissions is dictated by the science of astronomy, not by aesthetics, but from it Crotty has appropriated the contours of the work he shows in galleries and museums— diagrammatic, documentary, and reduced to essences. This focus on the science of what he does may at times belie the fact that he was educated as an artist, receiving a master's degree from the University of California, Irvine, in 1980, and that he has taught art off and on at UCLA. By no means unschooled in contemporary art practice,

Fig. 80
The artist clearing
brush in Malibu

theory, and tenets of social survival, nor disconnected from the art scene that bustles in nearby Santa Monica's Bergamot Station, Crotty nonetheless keeps a low profile on his isolated hilltop in Malibu.

Over the last several years, the body of work he has produced there has taken three principal forms: books, drawings on paper, and globes. The books, many of them oversized, as if to imply the scale of their subject matter, contain series after series of drawings, diagrams, and notations on various sections of stellar and solar space. Their reference is to the naturalist's tradition of note taking and notebook making, the meticulous gathering of facts and observations laid down in

Fig. 81
Russell Crotty
Atlas of Galactic and Globular Star Cluster Drawings, 1998
20-page book,
ink on paper
(right-page view);
51 x 102 in. (open)
Courtesy of Shoshana Wayne Gallery, Santa Monica, California

Fig. 82
Russell Crotty
Title page of Western Skies, Volume 1, 2000
Ink on paper;
28 x 30 in. (open)
Courtesy of Shoshana Wayne Gallery, Santa Monica, California

Fig. 83
Russell Crotty
Western Skies, Volume 1, 2000
22-page book,
ink on paper;
28 x 30 in. (open)
Courtesy of Shoshana Wayne Gallery, Santa Monica, California

organized fashion, to be preserved and revisited as needed. Bound in classical fashion, these books are an allusion to the last several hundred years of stargazing; to the efforts of well-known figures such as Copernicus, Galileo, and Edwin Hubble; and also to the many unknown and forgotten others whose personal quests down through the ages have contributed to the science of astronomy. These books can be regarded as homages that seek to connect the contemporary probing of outer space with the less sophisticated investigations that preceded it. But the books also, by virtue of the hand and eye skills involved, comment on the merit of firsthand over mediated experience and suggest, hopefully rather than cynically, that technology's omnipotence need not deter us from relishing the profundity of a personal encounter with nature.

The recent *NGC 7789, Galactic Cluster in Cassiopeia* (fig. 77) is perhaps the most reductive of Crotty's works on paper. A circle of sky forty-eight inches in diameter, it reveals a cluster of stars in its center, surrounded by an inky, seldom punctuated darkness. The black-and-white drawing is

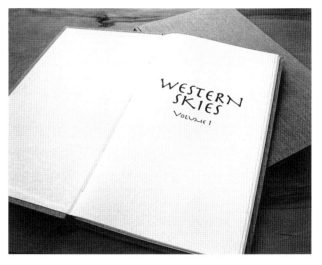

Fig. 84
Russell Crotty
Omega Centauri,
2000
Ink on paper,
mounted on
Lucite sphere;
diameter: 24 in.
Courtesy of Shoshana
Wayne Gallery,
Santa Monica,
California

rendered with hundreds of thousands of small marks made with ballpoint pen. Despite its occasional appearance through the telescope, color is absent from most of Crotty's oeuvre, banished for now to the realm of possibility in future works. Not unlike black-and-white photography—in which a reduction of forms to their essential, identifiable shapes in space seems to charge them, ironically, with almost a hyperreality—Crotty's diagrammatic drawings gain in authenticity and grandeur as they resist becoming colorful. Nothing perhaps could be more *essential* in its perspective than a view of the cosmos from trillions of miles away, and in the pursuit of essences, many twentieth-century artists—including Kazimir Malevich, Josef Albers, and Ad Reinhardt—turned to black-and-white. The simplified pictorial means that *NGC 7789* displays have evolved over a period of time, through experimentation with various media, culminating in Crotty's recent $900 purchase of this country's remaining stock of a particular archival India ink ballpoint pen—five hundred in all. Crotty's use of ballpoint began with his surfing drawings of a decade ago, and it permits a straightforward mark that carries little of the expressive baggage accompanying more traditional media. Further, ballpoint promotes the diagrammatic quality of scientific sketching, additionally confounding the question of whether he is acting as an artist or as an astronomer when making his work.

Crotty's planetarium-like globes (see figs. 75, 76, 84, 86) have enabled him to assume an even more complex perspective on human stargazing. Floating in the inner space of a gallery like so many planets and moons, the globes microcosmically become chunks of outer space matter, coalesced into orbs not unlike those of our own solar system. Each is rendered with a macrocosmic view of deep space, however, as if

Fig. 85
Russell Crotty
*Ghosts in the Void/
Planetary Nebular
Drawings*, 1998
Ink on paper,
bound in book;
37 x 37 in. (closed)
Private collection

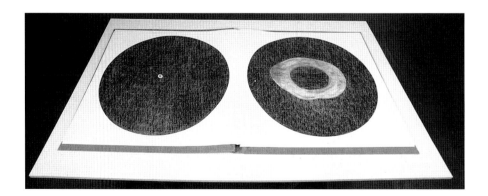

Fig. 86
Russell Crotty
NGC 2477,
Galactic Cluster
in Puppis, 2000
Ink on paper,
mounted on
Lucite sphere;
diameter: 24 in.
Private collection

Fig. 87
Russell Crotty
*Coastal Oaks and
the Cygnus Arm*,
2000
Ink and pencil on
paper; 48 x 48 in.
Courtesy of Shoshana
Wayne Gallery,
Santa Monica,
California

to say that within every planet, scavenged from the debris of cosmic origins and crafted into spheres of matter, are all the ingredients for an entire universe. In some cases the globes are specific locations in deep space, rendered in the round, which we spectators hurl around at the speed of light. Others are anchored to landscapes of the earth, shown in dark silhouette covering the globe's south pole, looking upward to sky in all directions north. By virtue of their three-dimensionality, the globes, which range from three to four feet in diameter, play with scale in an unnerving fashion, resonating with the incomprehensible distances of the stars rendered on their surfaces. At the telescope, Crotty sees in one eyeful a starscape trillions of miles across and sketches it onto a piece of paper that is so tiny in relation to what he is seeing as to be virtually nonexistent. Then those small sketches get "scaled up" to the size of the renderings on globes—a process that is absurd in its disparities, yet compellingly accurate in its portrayal of human audacity.

Significantly, the artist avoids appropriating space views from NASA photographs or utilizing high-tech telescopes that divert light around the naked eye for processing. These are boundaries beyond which he chooses not to go, for the same reasons that he surfs at night or in winter, only at places where he can isolate himself in the full experience of the activity. His work occupies an idiosyncratic position, referring to the virtues of a simpler technological past while at the same time pointing to the promise of our technologically advanced present. "The principal investigator of Solstice Peak Observatory" is how Crotty describes himself in a discussion about how his role straddles the boundary between art and science. From his observatory—in a way, his Walden Pond—the scale changes from city buildings and sidewalk perspectives to the curvature of the earth along the Pacific horizon, and there Crotty's dual role as artist and astronomer is a performance of sorts, an acting out of the proposition that science can be art, and art can be science.

Stephen Nowlin is the director of the Alyce de Roulet Williamson Gallery at Art Center College of Design, Pasadena, California.

Exploring *the* Solar System

With the advent of space exploration, bodies within the solar system that had been mere points of light were revealed as worlds of unexpected diversity. Even though surprisingly different in detail, these bodies have been shaped by many of the same basic processes. By examining commonalities and differences, we have gained a new appreciation for different styles of geological activity, the importance of collisional impacts, the complexity of weather systems, the prevalence of organic matter, and the potential for finding liquid water.

Geological Activity

On Earth, many features—such as mountain ranges, volcanoes, and earthquake faults—are associated with continental drift that is driven by an internal heat source. Although other bodies are shaped by similar processes, their geophysical evolution is often dramatically different.

Venus is Earth's twin, similar in size and density and orbiting the Sun at a comparable distance. There are, however, significant differences in the geophysical histories of the two planets. The surface of Venus is only about eight hundred million years old, much younger than that of Earth. Cloud-piercing radar images suggest that massive outpourings of lava from planet-wide volcanic eruptions may have been the cause. Two large volcanoes, Sif Mons and Gula (see fig. 89), are examples of such features.

Fig. 88
Voyager 1 looking back on Saturn and its rings
National Aeronautics and Space Administration, Jet Propulsion Laboratory, California Institute of Technology

Fig. 89
Three-dimensional
computer-generated
view of the surface of
Venus. Gula is three
kilometers (1.8 miles)
in height, and Sif
Mons (on left) has
a diameter of
300 kilometers
(186 miles) and a
height of 2 kilometers
(1.2 miles). The
simulated color
images taken by the
Magellan spacecraft
have a vertical
exaggeration of
22.5 times.
National Aeronautics
and Space
Administration,
Jet Propulsion
Laboratory,
California Institute
of Technology

Fig. 90
Active volcanic
eruption on Jupiter's
moon Io, imaged by
the *Galileo* spacecraft
NASA/JPL/University
of Arizona

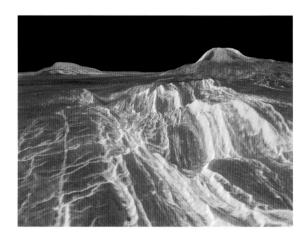

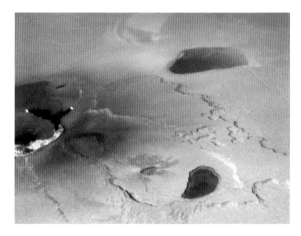

The distribution of volcanoes on Venus is also different. On Earth, volcanoes occur at boundaries where plates collide, such as the "Ring of Fire" around the Pacific Rim. By contrast, there appear to be no moving continental plates on Venus, and hundreds of thousands of volcanoes are distributed randomly around the planet. That two such similar planets have evolved so differently helps inform our understanding of the underlying geophysical processes.

We have also discovered that tidal heating can drive geophysical processes. Just as the gravitational pull of the moon creates ocean tides on Earth, Jupiter's immense gravity creates one-hundred-meter tides in the surface of its moon Io. This small moon (see fig. 90) has eight active volcanoes and more than one hundred hot spots—active volcanic areas glowing with lava flows. With one hundred times more volcanic activity than on Earth, Io is continually renewing its surface, erasing all record of its history.

As these comparisons of Earth, Venus, and Io illustrate, similar geophysical processes can alter the evolution of planets and moons in very different ways, resulting in dozens of surprisingly diverse bodies in the solar system.

Collisions

Although collisions might appear to result only in impact craters on a body's surface, we now realize that collisions can also greatly alter, often catastrophically, the evolution of planets and moons. The entire ring system of Saturn, for example, is likely the result of a breakup of one or more icy moons that were in orbit around the planet. Saturn's rings (see fig. 88) are made up of many small, icy particles a few centimeters in size; fewer larger ones; and fewer still of the largest ones, which are some meters across, as a result of a collisional process. Because of ongoing collisions in the ring system, the rings are not static but will continue to evolve.

Our own Moon likely resulted from the collision of a Mars-sized object with Earth. This object impacted early in the formation of Earth, melting the outer portion of our planet and splashing material into orbit, where it slowly condensed into the Moon. Much more recently, some sixty-five million years ago, a considerably smaller impact may have created an abrupt change in Earth's environment, causing the extinction of dinosaurs and many other species. Thus, collisions likely altered not only the geological evolution of Earth but the evolution of life on the planet as well.

Collisions continue to occur today. Comet Shoemaker-Levy, captured and fragmented by the strong gravitational forces of Jupiter, slammed into the planet in 1994 (see fig. 91). The collisions of the twenty-one observable fragments of the comet with Jupiter were the first such planetary impacts to be observed, and they illustrate the continuing role of collisions in affecting the evolution of bodies in the solar system.

Meteorology

Planetary weather systems such as Earth's are complex because they are shaped by many factors. Earth's weather is driven by solar energy and influenced, for example, by the oceans and continents and by a twenty-four-hour rotation. Other planetary atmospheres are shaped by different factors, providing us with new insight into meteorological processes.

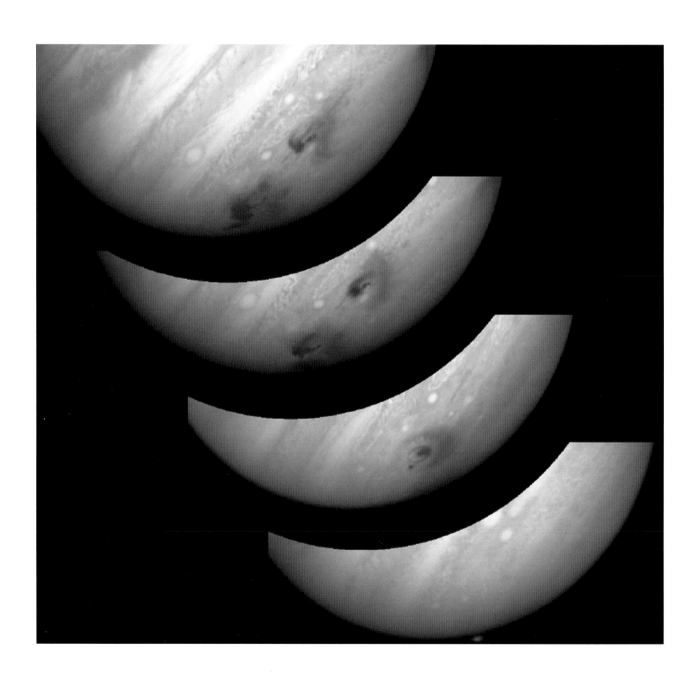

Fig. 91
Evolution of Comet Shoemaker-Levy 9
Fragment G impact site on Jupiter, imaged
by the Hubble Space Telescope
NASA/HST Comet Science Team

Fig. 92
Two of Jupiter's
moons, Io (left) and
Europa, shown by
Voyager 1 in front
of Jupiter's
Great Red Spot
National Aeronautics
and Space
Administration,
Jet Propulsion
Laboratory,
California Institute
of Technology

Jupiter is a giant, rapidly rotating sphere of gas and liquid, with no solid surface. Driven as much by internal heat from below as by sunlight from above, Jupiter's deep atmosphere has dozens of large, circular storms generated by the turbulent flow associated with jet streams of more than three hundred kilometers per hour. The Great Red Spot (see fig. 92) is a hurricane-like storm more than three Earth diameters across, which has persisted for at least four hundred years.

Six times farther from the Sun than Jupiter, Neptune has only 5 percent as much energy to drive its winds. Even so, its methane-rich blue atmosphere has jet streams of more than two thousand kilometers per hour, the fastest observed in the solar system. With less energy, there are few great storm systems, and turbulence (the likely cause of the slower winds on Jupiter) is greatly reduced. Even Neptune's Great Dark Spot (see fig. 93) is a transient feature, disappearing after only a few years.

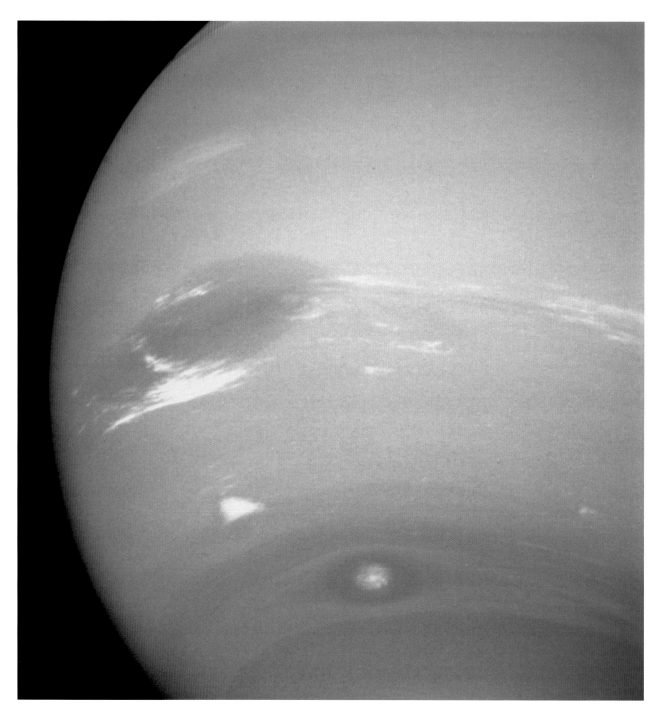

Fig. 93

Neptune's Great Dark Spot storm system accompanied by bright white clouds and another storm further south

National Aeronautics and Space Administration, Jet Propulsion Laboratory, California Institute of Technology

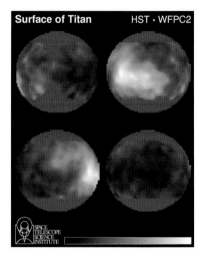

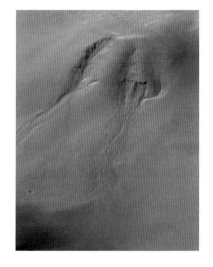

Although the same basic processes shape each weather system, their very notable differences provide important clues about the importance of various factors in creating such diverse atmospheric conditions.

Organic Materials

Complex chemical processes in the Earth's atmosphere and oceans produced prebiotic organic molecules, from which life is thought to have evolved. Although the Earth's warm and wet environment fosters such processes, organic molecules are also abundant in the cold, dry outer reaches of the solar system and in interstellar space beyond.

Comets are frozen ice and rock left over from when the solar system formed 4.6 billion years ago. Their surfaces are charcoal black, however, and they spew a fine dust of organic matter. A similar material darkens the icy surfaces of many of the moons and some of the ring particles in the colder regions of the solar system. During its earliest epoch, Earth was bombarded by comets, which may have contributed their organic material to the prebiotic soup out of which life evolved. As a result, there is particular interest in retrieving a sample of a comet surface so that we can understand its origin and its molecular constituents.

Saturn's moon Titan is a natural laboratory for studying the synthesis of organic molecules. As large as the planet Mercury, Titan has an atmosphere of mainly nitrogen, like Earth, but lacks the oxygen produced by living organisms. Instead, there is methane (natural gas), which is converted by solar and particle radiation into complex organic molecules, some forming an opaque haze. Using infrared rather than visible light, the Hubble Space

Fig. 94
First-time images of Titan's surface with prominent bright area 4,000 kilometers (2,500 miles) across, about the size of the continent of Australia
NASA/University of Arizona/Lunar and Planetary Laboratory

Fig. 95
High-resolution view from the *Mars Global Surveyor* spacecraft showing channels and associated aprons of debris that are interpreted to have been formed by groundwater seepage, surface runoff, and debris flow
NASA/JPL/Malin Science Systems, San Diego, California

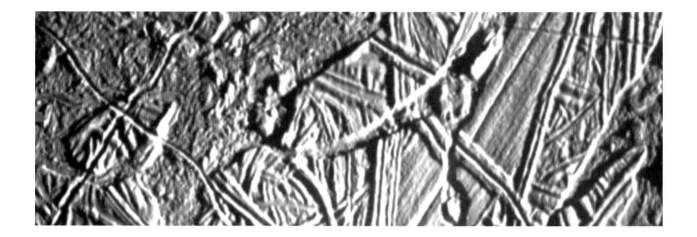

Fig. 96
Ice pack on Jupiter's
moon Europa.
NASA/JPL/University
of Arizona

Telescope peered through the haze, showing brighter and darker regions on the surface (see fig. 94). The darker areas may be lakes formed by liquid organics raining out of the atmosphere, while the brighter areas may be a solid surface coated with solid organic matter.

Although the surface of Titan is too cold for the complex chemistry that led to life on Earth, the chemistry occurring there today may provide important clues about the processes that occurred in early Earth's atmosphere before life evolved.

Liquid Water

Microbial life on Earth is remarkably robust, surviving in extreme environments such as freezing temperatures within Antarctic sea ice, near-boiling ocean vents, within rocks deep in the Earth's crust, and in acidic rivers. This suggests that life might have evolved elsewhere in the solar system where there was or is liquid water. For example, billions of years ago, great floods carved massive canyons in the surface of Mars. Evidence of gullies on Mars suggests that transient streams have more recently burst out of canyon walls (see fig. 95). Such gullies and dry lakebeds will likely be sites for future exploration.

Mars is not the only place to look for water. Tidal flexing at Europa's icy crust as it orbits Jupiter likely melts the ice, forming an ocean beneath the frozen surface. Close-up images reveal a cracked and ridged surface with reddish brown, salty deposits from below. In some areas the highly patterned, regular surface appears to have broken into ice rafts that moved about until the surface refroze (see fig. 96). The possibility of an ocean on Europa offers another prospect for the search for life elsewhere.

The first four decades of space exploration have revealed dozens of diverse worlds that beckon us to return for a closer look. Future missions will probe more deeply into the origins and evolution of these worlds by exploring on and beneath their surfaces (see fig. 97), analyzing samples that will further our insight into processes that have shaped these worlds and fostered the origin and evolution of life.

Edward C. Stone is director of the Jet Propulsion Laboratory, Pasadena, California.

Fig. 97
A 360-degree panorama imaged by the *Mars Pathfinder* spacecraft
National Aeronautics and Space Administration, Jet Propulsion Laboratory, California Institute of Technology

Humanity *and the* Universe
A Timeline

The Neolithic Age

The development of agriculture by tribes around the globe sparks the need to keep an accurate seasonal calendar, based on cycles of the Sun and Moon, for planting and harvest.

2600–2500 B.C.

The solar cult becomes the official court religion of Egypt. The Great Pyramid of Khufu is built in Giza with a portal facing the North Star to allow the dead pharaoh to reach the heavens.

2500

China adopts a calendar that marks 365 days in a year.

18th century

The Venus Tablet is inscribed under the reign of the Babylonian king Ammisaduqa. It records the position of the planet Venus over a twenty-one-year period.

14th century

The pharaoh Akhenaten attempts to place Egypt under a monotheistic belief system centered around the sun symbol Aten.

800

Babylonian astronomers record eclipses, planets in retrograde, and other celestial phenomena and predict such events with surprising accuracy. Their data influence Greek astronomy half a millennium later.

800–400

The Olmecs of La Venta, Mexico, develop a precise astronomical calendar that is used until the end of the Aztec civilization in the sixteenth century A.D.

Fig. 99
Porcelain Bowl
China, Qing dynasty (1644–1911)
Pacific Asia Museum Collection, Gift of Robert and Sheila Snukal, 1996

Fig. 98 (opposite)
Egg Nebula CRL 2688 seen from the Hubble Space Telescope,
January 6, 1996
NASA/Space Telescope Science Institute

Fig. 100
Siddhartha Meditating below Jambu Tree
Pakistan, Gandhara period, 3rd century
Schist; 23 x 17 x 5 (diam.) in.
The Norton Simon Foundation, Pasadena, California

6th century

Greek civilization embraces philosophy, which leads astronomers to shift their emphasis from merely predicting events to conceptualizing their cause. Parmenides is the first to conceive of the universe as a sphere and to conclude that moonlight is a reflection from the sun.

Greek astronomer Anaximander theorizes that the Earth is a cylinder suspended on a fixed point in space at an equal distance from all other heavenly bodies.

600

In Iran, Zoroaster proclaims a dualistic cosmology of the spirit in which good and evil are represented by light and darkness. This philosophy, known as Zoroastrianism, influences many future Judeo-Christian and Islamic cosmologies.

580–540

The Upanishads are written in India. Included are texts about the continual cycle of rebirth due to one's karmic actions (*samsara*).

Circa 550

Daoism is founded in China by Lao-tzu. The Dao describes the universe as composed of two principal forces: the active (yin) and the passive (yang).

Jainism is founded in India by Mahavira. Jains strive to release themselves from *samsara* (perpetual rebirth on earth) through strict ascetic practice and nonviolence to all living things.

Pythagoras, a Greek practitioner of the newly popular science of geometric astronomy, concludes that the Earth is spherical. Followers of Pythagoras determine that the universe operates on mathematical and geometric principles. They also conclude that each planet emits a sound, the tone being relative to its speed of revolution around the Earth, and that the universe therefore creates a musical harmony.

500–460

Buddhism is founded in India by Siddhartha Gautama. Buddhism emphasizes release from *samsara* by enlightenment through the practice of meditation.

5th to 3rd century

The Atomist group of Greek philosophers, notably Democritus and Epicurus, theorize that the universe is composed entirely of indivisible particles (*atoma*) in constant motion and inhabiting empty space. According to them, the attraction of these particles is the basis of every physical body.

4th century

The heavenly model accepted by Greek astronomers depicts the ball-shaped Earth sitting at a fixed point surrounded by a spinning spherical shell, with the stars implanted in the shell. The Sun, Moon, and planets revolve at independent speeds, and are therefore referred to as "wanderers."

Greek philosopher Plato explains that the independent revolutions of the Sun, Moon, and planets are due to their orbital placement between Earth and the stellar sphere.

Plato's former student Aristotle theorizes that stars travel through an invisible substance called ether and are propelled to move by desire.

Anaxagoras of Clazomenae theorizes that the Moon's topographic surface resembles the Earth's and explains the cause of solar eclipses.

2nd to 1st century

The Great Stupa, a dome-shaped reliquary monument to Buddha, is built at Sanchi, India. The stupa "egg" represents the physical universe united with the Buddha's nature and is pierced in the center by a rod (*axis mundi*) that links the earth to the heavens.

Fig. 101
The Ptolemaic Universe II of Robert Fludd The Henry E. Huntington Library and Art Gallery, San Marino, California

1st century B.C.

The astronomer Sosigenes of Alexandria, commissioned by Julius Caesar, devises the Julian calendar, which has a 365-day year and an extra day every four years (the "leap year"). This calendar is the basis for the modern Gregorian calendar.

1st century A.D.

The two largest monuments in Mexico, the Pyramid of the Sun and the Pyramid of the Moon, are constructed in Teotihuacan (northeast of modern Mexico City).

Greek biographer Plutarch writes "On the Face in the Moon," which debates the Aristotelian Stoic and Platonic Academics' views of the Moon's elemental composition and its relationship to Earth. In the dialogue are references to gravitational pull and centrifugal force—concepts not recognized until fifteen hundred years later.

60–100

Chinese astronomers record eclipses and observe planetary motions.

125

The Pantheon is built in Rome under the patronage of Emperor Hadrian. The domed temple, housing statues of the principal Roman gods—such as Jupiter, Mars, and Venus—becomes a symbol of the empire's dominion over most of the known world.

Circa 140

Greek astronomer Claudius Ptolemy publishes the *Almagest*, which creates a model of the solar system that places Earth at the center, with the Sun and all other planets revolving around it. Ptolemy's geocentric system is accepted in the West until the fifteenth century.

4th century

Chinese astronomers believe that the blue of the sky is an illusion and that the Sun, Moon, and stars float freely in space.

Circa 330

In China, Yu Xi is one of the first astronomers to describe the precession of the equinoxes.

5th century

The Chinese use water-driven armillary spheres, which revolve in phase with the stars.

476–550

Indian astronomer Aryabhata writes about the rotation of the Earth and mentions the epicyclic movement of the planets.

6th century

Italian bishop Isidorus of Seville is the first to draw the distinction between astrology and astronomy.

The cross-plan design of many medieval Christian cathedrals is situated so that the altar faces the east—the rising sun—to signify the resurrection of Christ.

Fig. 102
A British
reproduction of an
Arabic cosmological
astrolabe, c. 1880
Courtesy of the
Archives, California
Institute of
Technology,
Pasadena, California

700–730

Mayan science reaches its peak with detailed astronomical observations and the advanced use of mathematics.

800

The mystical Buddhist monument Borobudur is built in central Java, Indonesia, as a three-dimensional mandala.

833

An observatory is built in Baghdad. Arabia becomes the Near East's and Europe's leading center for astronomy, alchemy, and mathematics, based on concepts from ancient India, Greece, and Jewish and Muslim mysticism.

840

Al-Farghani writes the *Elements*, a summary of Ptolemy's *Almagest*. His translation contributes to the dominance of Ptolemaic astronomy in Europe until 1600.

1066

The sighting of Halley's Comet in England becomes astrologically linked to the murder of King Harold by William the Conqueror.

1100

The Arabs refine the astrolabe, a scientific instrument used to determine altitude, latitude, and points of the compass.

Circa 1200

The *I Ching*, an ancient Chinese text that presents a cosmology involving nature and humans in a single system, is recovered by Chu Hsi.

13th century

Jayavarman VII, ruler of Angkor (Cambodia), builds a temple whose architecture alludes to Mount Meru (the axis of the universe in Indian cosmology), where all life is born. At its gate stand a line of guardian-gods who use a snake to churn life from the "mountain-temple."

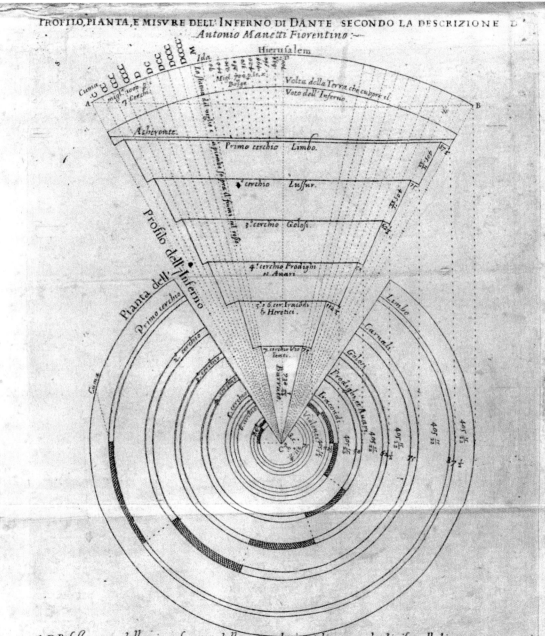

Fig. 103

Engraving showing the seven circles of hell from Dante's *Divine Comedy* (Florence, 1595)

Courtesy University of Notre Dame, Notre Dame, Indiana

1269

The second-earliest Japanese nativity horoscope is written. It includes the twelve Chinese directions, twelve signs of the zodiac, twenty-eight lunar mansions, planetary positions, and twelve houses. The horoscope reveals Chinese, Hellenistic, and Indian influences.

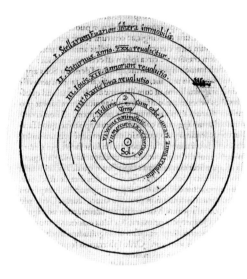

Fig. 104
Nicolaus Copernicus
*De Revolutionibus
Orbium Celestium*,
1543
The Henry E.
Huntington Library
and Art Gallery,
San Marino,
California

1413

A seasonal and astrological calendar book, the *Très riches heures du duc de Berry*, is illustrated by the Limbourg brothers in France.

Circa 1507

Italian Renaissance painter and scientist Leonardo da Vinci composes notes and illustrations on subjects such as physics, astronomy, mechanics, and geology. In the manuscript later known as the Codex Leicester, da Vinci suggests that moonlight is caused by the sun's reflection on the moon's waters.

1543

Rejecting Ptolemy's complicated theory of earth-centered planetary motion, Polish astronomer Nicolaus Copernicus writes *De revolutionibus orbium coelestium libri vi*, which describes all planets as revolving around the sun and clarifies phenomena such as retrograde motion.

1572

Danish astronomer Tycho Brahe discovers a new "star" (supernova) in the constellation Cassiopeia. This discovery, recorded in his book *De nova stella* in 1573, contradicts the prevailing belief that the stars are perfect and unchanging.

17th century

French inventor and philosopher Blaise Pascal writes his *Pensées*, which challenges the reader to imagine the universe as infinite space, not only on an immense scale but also in a microcosmic sense. The work disrupts the Christian notion of a world created by God for humankind.

1608

Hans Lippershey develops a telescope.

1609

Galileo Galilei builds a telescope, and through it he discovers mountains, valleys, and craters on the moon as well as dark blemishes on the sun (sunspots).

Johannes Kepler, using Brahe's observations, deduces that planetary orbits are elliptical, not circular, with the sun at one focus of each ellipse. Two of his three laws of planetary motion are published in his book *Astronomia nova*.

1610

Galileo discovers three satellites revolving around Jupiter, thereby shattering the common notion of a geocentric universe.

Galileo's pupil Benedetto Castelli develops a safe method for sun observation by directing the telescope's eyepiece to a screen.

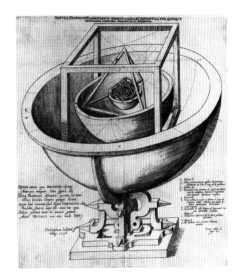

Fig. 105
Johannes Kepler
Mysterium Cosmographicum,
1597 (1621 ed.)
Courtesy of the Archives, California Institute of Technology, Pasadena, California

1617

The Neoplatonist physician Robert Fludd writes *Utriusque cosmi historae*, stressing the divine proportional relationships between the cosmos, the terrestrial elements, and music.

Fig. 106

Jan van der Heyden, *Library Interior with Still Life*, 1711–12; oil on canvas, 27 x 22 ½ in.

The Norton Simon Museum, Pasadena, California

1630

Galileo completes his *Dialogue Concerning the Two Chief World Systems, Ptolemaic and Copernican*. The book is published in 1632, and Pope Urban VIII, believing it to be ridiculing him, condemns Galileo for heresy.

1644

Philosopher René Descartes publishes his *Principia Philosophiae*. In it he rejects the idea of the heavens as a vacuum and instead envisions watery current as a vehicle for motion in the universe.

1659

Christiaan Huygens constructs the first maps of Mars based on telescopic observations.

1687

Isaac Newton defines gravity in his *Principia* in terms of relationships between objects due to their differences in mass. The book explains centripetal force, which propels a planet into orbit.

18th century

Sawai Jai Singh of Jaipur, India, constructs five stone observatories on the Subcontinent to make accurate astrological tables. The Jaipur Observatory includes the largest sundial in the world, whose projecting arm is ninety feet high.

1761–69

Transits of Venus during these two years allow scientists to compute the size of the solar system.

1781

English astronomer William Herschel discovers a distant "star," which he names Georgium Sidus, after King George III. The star is now known as the planet Uranus.

1838

John Herschel, son of royal astronomer William Herschel, travels to Cape Town, South Africa, and creates the most complete telescopic map of the Southern Hemisphere of its time.

1846

The planet Neptune is discovered by French mathematician U. J. J. Leverrier and Berlin Observatory assistant Johann Gottfried Galle.

1846–50

English amateur astronomer William Lassell discovers moons orbiting Uranus, Neptune, and Saturn.

1855

The first significant photograph of the sun is shot by physicists Armand Fizeau and Jean-Bernard-Léon Foucault.

Fig. 107
Sketch for the
Georges Méliès film
*A Trip to the
Moon* (1902)
Courtesy of the
Academy of
Motion Picture Arts
and Sciences,
Beverly Hills,
California

1883

English amateur astronomer Ainslie
Common makes the first successful
photograph of the Orion Nebula with
the assistance of Edison's newly
invented electric light.

1884

New scientific discoveries regarding the
nature and composition of light inspire
neo-impressionist painter Georges
Seurat to adapt the method of "color
division" (pointillism) to depict light
and color relationships.

1900

Max Planck develops the quantum
theory of physics.

1902

Georges Méliès produces the science
fantasy silent film *A Trip to the Moon.*

1903

The first successful airplane is invented
and flown by Orville and Wilbur
Wright in Kitty Hawk, North Carolina.

1904

George Ellery Hale founds Mount Wilson Observatory near Pasadena, California.

1907

Pablo Picasso paints *Les demoiselles d'Avignon*, a forerunner of the cubist movement.

1909

A group of iconoclastic Italian painters, including Giacomo Balla and Umberto Boccioni, write the *Manifesto of Futurist Painters* and create works depicting motion and employing technical elements derived from the work of the cubists and Paul Cézanne.

1915

Albert Einstein publishes a visual model of the theory of relativity that radically alters Western civilization's view of its world.

Robert Goddard, a physics professor at Clark University in Worcester, Massachusetts, proves that a rocket can propel itself in a vacuum, making space travel a distant possibility.

1922–24

Edwin Hubble, using the Hooker Telescope (the world's most powerful telescope) at Mount Wilson Observatory, discovers galaxies outside the Milky Way, which was previously believed to make up the entire universe.

1926

Goddard launches the first liquid-fueled rocket to a height of forty-one feet.

1927

A practitioner of a mode of abstract painting that stresses nature's inherent simplicity and geometry, Wassily Kandinsky composes the planetary painting *Heavy Circles*.

1929

Hubble makes further breakthroughs concerning the size and nature of outer space, including the idea that the universe had a beginning and is continually expanding.

1930

Amateur astronomer Clyde W. Tambaugh discovers the planet Pluto at the Lowell Observatory at Flagstaff, Arizona.

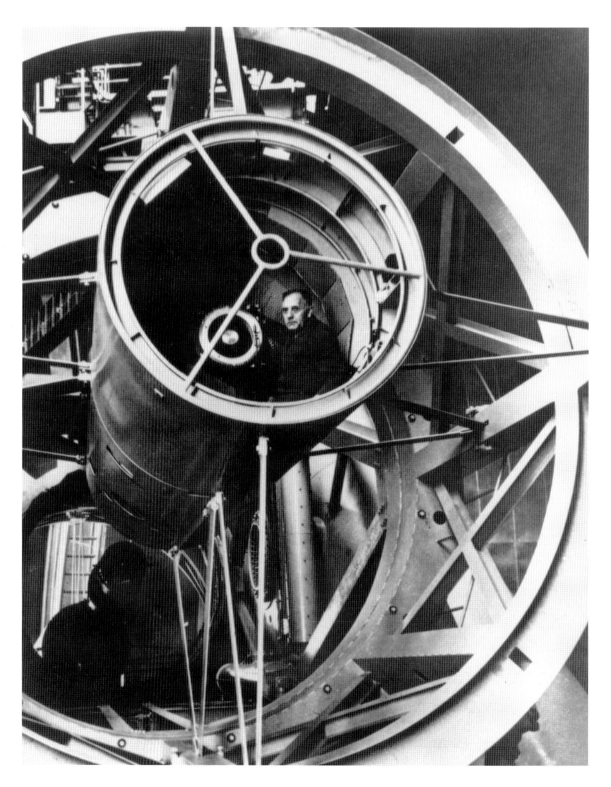

Fig. 108

Astronomer Edwin Hubble inside the Mount Wilson telescope

The Carnegie Institution of Washington

1951

Responding to the Cold War, American filmmaker Robert Wise directs *The Day the Earth Stood Still*, a film about the human response to contact with visitors from another planet.

1957

The USSR conducts the first-ever space flight with the launch of the satellite *Sputnik*.

1961

Russian cosmonaut Yuri Gagarin is the first man in space.

1966

The television series *Star Trek* begins.

1969

As part of the *Apollo 11* mission, American astronauts Neil Armstrong and Edwin E. Aldrin Jr. are the first humans to step on the Moon's surface, an event that is televised around the world.

Stanley Kubrick directs the film *2001: A Space Odyssey*, based on a science-fiction novel by Arthur C. Clarke.

American mixed-media artist Robert Rauschenberg prints the Stoned Moon series of lithographs, inspired by the Apollo space missions.

1970

The first satellite X-ray observatory, *Uhuru*, is launched.

1972

The stereoscopic ultraviolet satellite telescope *Copernicus* is launched to observe stars and interstellar gas.

1976–77

The *Viking* space probes reach Mars, and the resulting photographs construct the most detailed topographic maps of the planet to date.

Fig. 109
Alexander Calder
Maelstrom with Blue,
1967
Gouache on paper;
43 x 29 1/4 in.
Norton Simon
Museum, Pasadena,
California,
Gift of Mr.
W. H. Hal Hinkle,
New York, 1986

1977

Rings are discovered around the planet Uranus.

1978

The satellite observatory *Einstein*, using ultraviolet high-resolution imaging, is launched.

1979

Space probe *Voyager 1* detects the first known ring around Jupiter.

1980

Voyager 1 reveals the complex structure of Saturn's rings.

1988

Stephen Hawking publishes *A Brief History of Time: From the Big Bang to Black Holes*, suggesting that the universe paradoxically has a boundary and no boundary. His discussion reflects humankind's increasingly blurring concept of a concrete reality.

1990

The Hubble Space Telescope is placed in orbit around the Earth and soon begins imaging distant sections of the universe.

1992

George Smoot and a team of scientists, using the Cosmic Background Explorer satellite, are able to detect "static" left over from the "big bang" and announce their theory of how the universe was born.

1995

A NASA space probe enters Jupiter's atmosphere.

1997

The motion picture *Contact*, directed by Robert Zemeckis, is released.

Mars Pathfinder lands on the red planet and captures the most sophisticated panoramic photographs of the planet's surface to date.

Fig. 110
Galaxy Cluster MS1054-03 seen from the Hubble Space Telescope, July 15, 1999
NASA/Space Telescope Science Institute/ESA

Selected Bibliography

Danielson, Dennis Richard. *The Book of the Cosmos: Imagining the Universe from Heraclitus to Hawking*. Cambridge, Mass.: Helix Books, 2000.

Eames, Charles and Ray. Timeline for *Mathematica: A World of Numbers and Beyond*. Provided by the California Science Center.

Holborn, Mark. "The Map." In *Steps in Space*, special issue of *Aperture*, no. 157 (fall 1999): 2–9.

Knight, Christopher. "The Persistent Observer," in *Twenty-four Years of Space Photography*. Pasadena, Calif.: Baxter Art Gallery, California Institute of Technology; New York: W. W. Norton, 1985.

Jennifer Gunlock is gallery assistant at the Armory Center for the Arts.

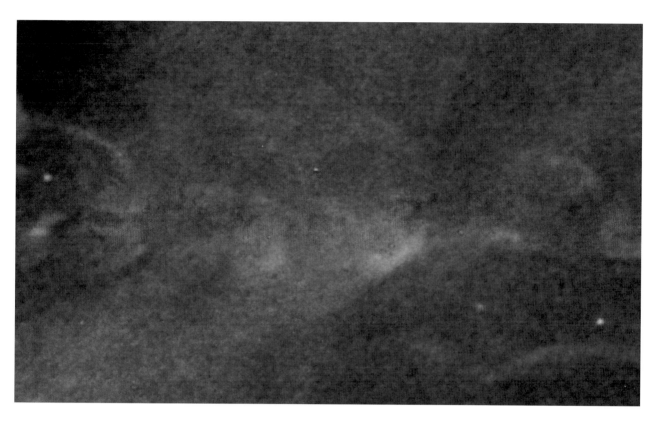

Above and overleaf:
Jets from young stars
seen from the Hubble
Space Telescope,
July 6, 1995
NASA/Space
Telescope Science
Institute

Participating Institutions

The Alyce de Roulet Williamson Gallery, located on the campus of Art Center College of Design, hosts exhibitions of contemporary art and design, with an emphasis on new media.

The Armory Center for the Arts builds on the power of art to transform lives by engaging the community through the creation, presentation, and teaching of the visual arts. The Armory's innovative program strategies serve as models for the fields of arts and education.

California Institute of Technology Public Events provides a broad spectrum of multidisciplinary cultural, information, and entertainment programs. These programs are designed to complement the experience of Caltech students, faculty, and staff, as well as to provide outreach and cultural enhancement for the community.

The Huntington Library, Art Collections, and Botanical Gardens is a research and cultural center with world-renowned collections of rare books, manuscripts, and artworks.

The Norton Simon Museum is regarded as one of the most remarkable art collections ever assembled. Seven centuries of European art from the Renaissance to the twentieth century are on view, including works by such renowned artists as van Gogh, Picasso, Raphael, Rembrandt, and Zurbarán. The museum also boasts outstanding Asian sculpture from India and Southeast Asia, as well as postwar American art.

One Colorado is an outdoor marketplace where cultural experiences are an integral part of shopping at Armani Exchange or dining at Il Fornaio. By fusing historic architecture and modern boutiques and restaurants with the third dimension of culture, One Colorado's position as a vibrant meeting place is used to touch a diverse audience with meaningful programming.

The Pacific Asia Museum, housed in a historic Chinese-style building, specializes in the art and cultures of Asia and the Pacific Islands, informing the public about the rich heritages of these diverse cultures through its collections, exhibitions, cultural festivals, and other educational programs.

Founded in 1987, Southwest Chamber Music is one of the most active chamber music ensembles in the United States, presenting concert series at the Norton Simon Museum in Pasadena, Herbert Zipper Concert Hall at the Colburn School of Performing Arts in Los Angeles, and a celebrated summer season at the Huntington Library, Art Collections, and Botanical Gardens. The ensemble offers Open Rehearsal: Breaking the Code, one of the few regularly scheduled open rehearsal series in the nation, at the Armory Center for the Arts.

Photography Credits

Most of the photographs in this volume have been supplied by the owners or custodians, as identified in the captions. The following additional photography credits apply to the page numbers listed.